Jennifer Kyrnin

Sams **Teach Yourself**

Bootstrap

in **24**
Hours

SAMS 800 East 96t

Sams Teach Yourself Bootstrap in 24 Hours

ISBN-13: 978-0-672-33704-8

ISBN-10: 0-672-33704-5

Library of Congress Control Number: 2015914168

Printed in the United States of America

First Printing November 2015

Trademarks

Warning and Disclaimer

Special Sales

For information about buying this title in bulk quantities, or for special sales opportunities (which may include electronic versions; custom cover designs; and content particular to your business, training goals, marketing focus, or branding interests), please contact our corporate sales department at corpsales@pearsoned.com or (800) 382-3419.

For government sales inquiries, please contact governmentsales@pearsoned.com.

For questions about sales outside the U.S., please contact international@pearsoned.com.

Acquisitions Editor
Mark Taber

Managing Editor
Sandra Schroeder

Senior Project Editor
Tonya Simpson

Copy Editor
Megan Wade-Taxter

Indexer
Tim Wright

Proofreader
Chuck Hutchinson

Technical Editor
Jon Morin

Editorial Assistant
Vanessa Evans

Cover Designer
Mark Shirar

Compositor
codeMantra

Contents at a Glance

Part IV: Customizing Bootstrap

Table of Contents

About the Author

Jennifer Kyrnin has been teaching HTML, XML, and Web design online since 1997. She has built and maintained websites of all sizes from small, single-page brochure sites to large, million-page databased sites for international audiences. She focuses on responsive design using Bootstrap and WordPress. She lives with her husband and son and numerous animals on a small farm in Washington state.

Dedication

As always, to Mark and Jaryth. I love you.

Acknowledgments

I would like to thank all the people at Pearson for the opportunity to write this book and work with you. I would particularly like to thank Mark Taber for understanding and helping with issues that came up, as well as my wonderful tech editor Jon Morin and my copy editor Megan Wade-Taxter for all the great suggestions and corrections. Any technical errors you find in the book are mine alone; they probably tried to stop me.

I would like to thank the many and varied Bootstrap designers out there who helped me either knowingly or unknowingly with their amazing designs and wonderful ideas. I also need to thank my writing group—Jerry, Karen, Ted, Renee, and Rob—for letting me inject nonfiction into our regular fiction discussions.

And as usual, I couldn't have completed this book without the help of my family.

We Want to Hear from You!

As the reader of this book, *you* are our most important critic and commentator. We value your opinion and want to know what we're doing right, what we could do better, what areas you'd like to see us publish in, and any other words of wisdom you're willing to pass our way.

We welcome your comments. You can email or write to let us know what you did or didn't like about this book—as well as what we can do to make our books better.

Please note that we cannot help you with technical problems related to the topic of this book.

When you write, please be sure to include this book's title and author as well as your name and email address. We will carefully review your comments and share them with the author and editors who worked on the book.

Email: feedback@samspublishing.com

Mail: Sams Publishing
        ATTN: Reader Feedback
        800 East 96th Street
        Indianapolis, IN 46240 USA

Reader Services

Visit our website and register this book at informit.com/register for convenient access to any updates, downloads, or errata that might be available for this book.

Introduction

Bootstrap is a web design framework that makes it easy for you to build responsive websites quickly and effectively. Web developers at Twitter created Bootstrap to help them build their website. They released it as an open source framework in 2011, and it has become one of the most popular frameworks on the Web.

NOTE

Visit our website and register this book at **informit.com/register** for convenient access to any updates, downloads, or errata that might be available for this book.

Bootstrap Makes Building a Complex Site Easy

This book covers the basics of Bootstrap. But it also explains how to use Bootstrap to enhance your site and add features that many designers would have otherwise ignored. Once you've finished this book, you will understand:

- ▶ How to use grids to create a beautiful site layout

- ▶ The use of labels, badges, panels, and wells to add interest to your text

- ▶ How to style tables and forms so that they are easy to read, look nice, and are responsive

- ▶ Using images on Bootstrap pages and how to add the icons included in the Bootstrap framework

- ▶ How to quickly create navigation and buttons that use features like dropdown menus, search fields, and multiple menu levels

- ▶ The CSS utilities Bootstrap provides to add features such as alignment, color, and visibility

- ▶ How to use the JavaScript plugins provided by Bootstrap, including modals, tooltips, alerts, accordions, and image carousels

- ▶ The options you have for customizing Bootstrap, including CSS, Less, and Sass

- ▶ How to make Bootstrap sites accessible

- ▶ And where to go to learn even more about Bootstrap

How to Use This Book

This book is divided into 24 lessons, called "hours." Each lesson covers a specific topic related to building responsive web pages using responsive web design. Each lesson takes about an hour to complete.

Organization of This Book

The book is divided into four sections:

▶ Part I, "Getting Started with Bootstrap," introduces you to Bootstrap and web frameworks. You will learn the basics of frameworks, how to install Bootstrap, and how to build and use Bootstrap for a website.

▶ Part II, "Building and Managing Web Pages with Bootstrap," explains how to use the Bootstrap CSS styles and components to create websites.

▶ Part III, "Bootstrap JavaScript Plugins," teaches you how to use the Bootstrap JavaScript plugins to add features to your websites.

▶ Part IV, "Customizing Bootstrap," introduces the advanced features of Bootstrap web development to create complex designs that don't necessarily look like Bootstrap designs.

Conventions Used in This Book

Code samples are written in monospaced font within the text of the book, while blocks of code will be called out separately, for example:

```
This is a block
Of code
```

Some code samples that are too long to display as one line in the book use the ➥ symbol to indicate that these lines should be all on one line, like this:

```
<link rel="stylesheet" href="styles-320.css"
➥ media="only screen and (max-width:320px)">
```

This book has two types of sidebars:

NOTE

Notes provide additional information about the topics that are discussed in the hour. They may also provide interesting facts or tidbits about the related content.

CAUTION

Cautions alert you of things that can cause problems for your Bootstrap web designs.

You can also use the Try It Yourself sections to help you practice what you've learned in the hour.

TRY IT YOURSELF ▼

Nearly every hour will have at least one step-by-step tutorial called "Try It Yourself" to help you use what you've learned.

Q&A, Quiz, and Exercises

Every hour ends with a short question-and-answer section to help with follow-up questions that occur as a result of reading the hour. You can also take a short quiz on the hour as well as do some suggested exercises to help you get more out of what you learned and apply this knowledge to your own web designs.

Where to Go to Learn More

More information is available on the companion website, http://html5in24hours.com/bootstrapbook/, where you can go to see the examples, view and download the source code, view and report errata about the book, and continue to learn and ask questions about Bootstrap. You can also find Jennifer Kyrnin online at http://htmljenn.com/, and she welcomes questions and comments.

HOUR 1
What Is Bootstrap, and Why You Should Use It

What You'll Learn in This Hour:

▶ What a web framework is
▶ How web frameworks are used
▶ Some common web frameworks
▶ What Bootstrap is
▶ Pros and cons for using Bootstrap

Bootstrap is a web framework developed by Twitter to help make web pages and applications quicker to design. Because you've picked up this book, chances are you are interested in using Bootstrap for your web pages, but it is a huge tool with a lot of features, and it can be daunting to get started.

In this hour you will learn more about what a web framework is and how they help web developers build faster and more cost-effective websites. You will also learn how Bootstrap differs from some of the other web frameworks available for use. Finally, you will discover some of the reasons frameworks are a good idea for many websites and why Bootstrap might be the perfect solution.

What Is a Web Framework?

To understand Bootstrap, you first have to understand web frameworks. A web framework is a tool that programmers and web developers can use to simplify a complex system, such as a website or web application. A web framework is a development framework for websites, and several are available for designers and developers to use, including

▶ Foundation (http://foundation.zurb.com/)

▶ Pure CSS (http://purecss.io/)

▶ HTML5 Boilerplate (http://html5boilerplate.com/)

▶ Responsive Grid System (http://www.responsivegridsystem.com/)

And, of course, Bootstrap. All these frameworks provide HTML, CSS, and sometimes JavaScript tools to the developer to provide the underlying structure and functions of a website.

Most web design frameworks include a layout or grid system that makes it easy to create multi-column sites quickly. The best frameworks also include CSS to style tables, manage forms, create buttons, style typography, and they are responsive.

Bootstrap is a web framework that offers all these features and more.

A Framework Is More Than Just a Template

A framework for web pages is more than just a template or even a series of templates. Instead, it is a group of tools you can use to create your web pages.

Many people use frameworks as if they were templates, and that is a great way to get started using them. But you aren't getting the most out of your framework if all you do is create standardized pages based on the sample templates you can find.

A framework helps you manage your web pages by taking care of the tedious, repetitive tasks ahead of time so that you can focus on the actual design.

One of the most difficult things to create is a cohesive layout that stays consistent throughout your designs. Creating an effective grid layout means doing a lot of math. Every time you add another column to your grid, you have to calculate the gutters between margins, the change in column widths, and how they flow together across the entire page. This is especially difficult when building a responsive website because then you have to do all that math two to three times—once for each layout. Most designers create their grid layout page by page. And this means that you end up with almost random collections of columns and rows throughout the site.

A framework has all this figured out for you. You'll learn more about Bootstrap grids in Hour 5, "Grids and How to Use Them," but Figure 1.1 shows an example of how Bootstrap handles layout grids.

As you can see in Figure 1.1, Bootstrap uses a 12-column default grid, and you can divide it into many different sets of columns just by using some HTML classes on your elements. Without the framework, you would have to build all those classes and their relationships to each other by hand. You can see the HTML that generates that grid in Listing 1.1.

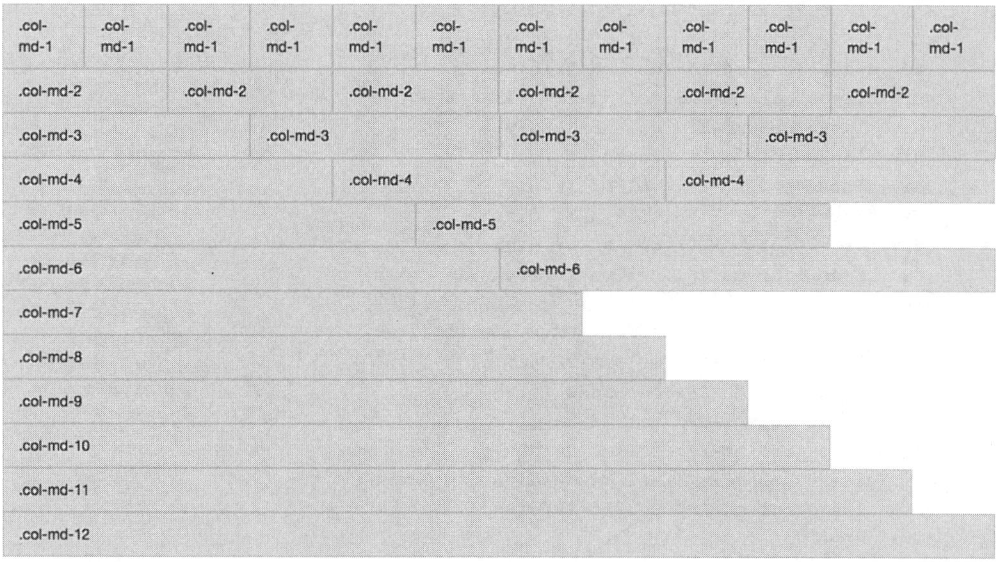

.col-md-1	.col-md-1	.col-md-1	.col-md-1	.col-md-1	.col-md-1	.col-md-1	.col-md-1	.col-md-1	.col-md-1	.col-md-1	.col-md-1

FIGURE 1.1
The default Bootstrap grid.

LISTING 1.1 Creating a Grid System in Bootstrap

```
<!doctype html>
<html lang="en">
<head>
  <meta charset="UTF-8">
  <meta http-equiv="X-UA-Compatible" content="IE=edge">
  <meta name="viewport" content="width=device-width, initial-scale=1">
  <title>Basic Grid System</title>
  <link href="css/bootstrap.min.css" rel="stylesheet">
  <style>
  .show-grid [class^=col-] {
    padding-top: 10px;
    padding-bottom: 10px;
    background-color: #eee;
    background-color: rgba(86,61,124,.15);
    border: 1px solid #ddd;
    border: 1px solid rgba(86,61,124,.2);
  }
  </style>
</head>
<body>
  <div class="container">
  <h1>Basic Grid System</h1>
    <div class="row show-grid">
      <div class="col-md-1">.col-md-1</div>
      <div class="col-md-1">.col-md-1</div>
```

```
    <div class="col-md-1">.col-md-1</div>
    <div class="col-md-1">.col-md-1</div>
    <div class="col-md-1">.col-md-1</div>
    <div class="col-md-1">.col-md-1</div>
    <div class="col-md-1">.col-md-1</div>
    <div class="col-md-1">.col-md-1</div>
    <div class="col-md-1">.col-md-1</div>
    <div class="col-md-1">.col-md-1</div>
    <div class="col-md-1">.col-md-1</div>
    <div class="col-md-1">.col-md-1</div>
</div>
<div class="row show-grid">
    <div class="col-md-2">.col-md-2</div>
    <div class="col-md-2">.col-md-2</div>
    <div class="col-md-2">.col-md-2</div>
    <div class="col-md-2">.col-md-2</div>
    <div class="col-md-2">.col-md-2</div>
    <div class="col-md-2">.col-md-2</div>
</div>
<div class="row show-grid">
    <div class="col-md-3">.col-md-3</div>
    <div class="col-md-3">.col-md-3</div>
    <div class="col-md-3">.col-md-3</div>
    <div class="col-md-3">.col-md-3</div>
</div>
<div class="row show-grid">
    <div class="col-md-4">.col-md-4</div>
    <div class="col-md-4">.col-md-4</div>
    <div class="col-md-4">.col-md-4</div>
</div>
<div class="row show-grid">
    <div class="col-md-5">.col-md-5</div>
    <div class="col-md-5">.col-md-5</div>
</div>
<div class="row show-grid">
    <div class="col-md-6">.col-md-6</div>
    <div class="col-md-6">.col-md-6</div>
</div>
<div class="row show-grid">
    <div class="col-md-7">.col-md-7</div>
</div>
<div class="row show-grid">
    <div class="col-md-8">.col-md-8</div>
</div>
<div class="row show-grid">
    <div class="col-md-9">.col-md-9</div>
</div>
<div class="row show-grid">
    <div class="col-md-10">.col-md-10</div>
</div>
```

```
      <div class="row show-grid">
        <div class="col-md-11">.col-md-11</div>
      </div>
      <div class="row show-grid">
        <div class="col-md-12">.col-md-12</div>
      </div>
    </div>
  </body>
</html>
```

In Figure 1.2, you see the same HTML displayed on an iPhone. It's much narrower than the web browser view, but Bootstrap gives a nice display without any change to the code.

FIGURE 1.2
The default Bootstrap grid as shown on an iPhone.

You will learn more about creating grids with Bootstrap in Hour 5, but this gives you a quick taste for how Bootstrap works.

Pros and Cons of Frameworks

A web framework is a tool like any other, and there are definite benefits to using them, but there are also drawbacks. Some of the best reasons for using a framework include

▶ **Speed and efficiency**—As I mentioned previously, frameworks do a lot of the more common tasks for you so you can focus on the more important parts of your design. For example, if you're going to create a three-column section of your site in Bootstrap, you just need to know a few classes to create it.

▶ **Ease of use**—Most web frameworks offer features like typography that novice designers might ignore because they can be difficult. But with a framework, you can create a site that has top-notch typography without having to worry about the difference between ems and rems or how tall the line-height should be.

▶ **Maintainability**—When you use a framework for a website, you make it more maintainable in the future because the styles and scripts are, for the most part, predefined. If you build the site and someone else needs to maintain it, he doesn't have to learn as much about how you write CSS because most of that is handled by the framework.

▶ **Stability and security**—All the web frameworks I've mentioned are used by thousands of people around the world. They are well tested and maintained, and problems are found and fixed regularly. You don't have to worry as much about browser support or whether the designs will break because many people are testing them in more scenarios than most web designers can afford to test with on their own.

But, of course, there are also drawbacks to using a web framework:

▶ **Code bloat**—Although web frameworks are built to be as small and streamlined as possible, there is always going to be more there than you really need. In Hour 23, "Using Less and Sass with Bootstrap," I give you some suggestions for how to customize your Bootstrap installation to make it suit your needs.

▶ **Repetition**—It can sometimes be challenging to create a website using a framework that doesn't look like every other website using that framework. Throughout this book, I provide ideas for how to customize your site designs beyond the Bootstrap defaults.

▶ **Steep learning curve**—Frameworks are big and have a lot of things to learn to use well, and this can take time. But luckily you've purchased this book, so you're well on the way to solving this problem.

▶ **Less control**—Ultimately, when you use a framework you are giving up some control over how your website is built. If the framework designers use pixels for font sizes and you hate that, you have to either let your emotion go or manually go through and adjust it as you build your pages.

Every website is different, and it's important that you evaluate the needs of the site before trying to shoehorn it into a framework. But frameworks do add a lot of value, so it's always a good idea to consider them.

What Is Bootstrap?

Bootstrap is a web framework to help web designers and developers create websites and web applications. It is sometimes called "Twitter Bootstrap" because it was developed by Mark Otto and Jacob Thornton at Twitter to encourage more consistency in their internal tools and web applications.

It uses HTML and CSS for templates, typography, forms, navigation, buttons, tables, and more. And Bootstrap includes a JavaScript library for building dynamic page elements that are found on many modern websites. Bootstrap also includes a license to use a font library called Glyphicons to add graphical elements to your web pages quickly and easily.

How Is Bootstrap Different from Other Frameworks?

In general, most web development frameworks are the same. They offer CSS and sometimes JavaScript. They usually provide a grid system or some other way of laying out web pages. So, how do you choose between Bootstrap and some of the other options out there?

Some things you should consider include

▶ **Do you need software under a specific license?**
The license the framework is released under can affect how it may be used. Bootstrap is released under the MIT open source license.

▶ **Do you need specific technology like Less, Sass, or jQuery?**
Not all frameworks use these technologies, but Bootstrap offers all three.

▶ **Do you need a grid structure for your layouts, and do you need a responsive website?**
Most web frameworks start with a basic grid system, but some do not. And while responsive web design (RWD) is more common now, not all frameworks offer it. Bootstrap offers both RWD and a robust grid system.

▶ **Do you need to support legacy systems like Internet Explorer 8 and lower?**
Internet Explorer 8 handles HTML5 and CSS3 differently than more standards-compliant browsers, but not all frameworks take this into account. Bootstrap does.

▶ **Do you need typography included?**
You can always do the typography yourself, but many web frameworks come with a pre-built typography so you don't have to worry about line spacing and font sizes. Bootstrap comes with basic typography.

▶ **Do you need icons or buttons?**
Buttons make your site more interactive, and icons keep the site consistent looking. Bootstrap supports several types of buttons and is the only web framework to offer Glyphicons included in the license (with attribution).

▶ **Do you need support for tables or forms?**
If you're going to use more advanced HTML features like tables and forms, it's nice to have a framework that can support them. Bootstrap supports both tables and forms.

▶ **Do you need your site to be extremely small to minimize bandwidth use (such as for mobile-only sites)?**
One problem with many web frameworks is that they are huge and require a lot of bandwidth to download. Although Bootstrap is large at 150KB for the full (minified) installation, you can customize it to include only the features you need and reduce the size significantly.

The previous information is useful for knowing about Bootstrap, but it's also nice to know a little more about other web frameworks you can use. Table 1.1 shows you the features I listed here and how the different frameworks compare.

TABLE 1.1   Comparison of Different Web Frameworks

Feature	Bootstrap	Foundation 5	Pure CSS	HTML5 Boilerplate	Responsive Grid System
RWD	✓	✓	✓	✓	✓
Mobile first	✓	✓		✓	
Less	✓				
Sass	✓	✓			
jQuery and plugins	✓	✓		✓	
Grid system	✓	✓	✓		✓
Typography	✓	✓		✓	
Tables	✓	✓	✓		

Feature	Bootstrap	Foundation 5	Pure CSS	HTML5 Boilerplate	Responsive Grid System
Forms	✓	✓	✓		
Icons	✓				
Navigation	✓	✓	✓		
Internet Explorer 8 support	✓			✓	
Customizable	✓	✓	✓		✓
Buttons	✓	✓	✓		
Size	150KB	350KB	4.4KB	17KB	20KB
License	MIT	MIT	Yahoo BSD	MIT	Creative Commons 3.0 Attribution

All sizes are based on downloading the minified version of the CSS and JavaScript for the complete package—except for Pure CSS, which lists the size they claim after gzipping.

Why You Should Use Bootstrap

The choice of which web framework to use is an important one to make. But here are some of the reasons that I choose to use Bootstrap over other web frameworks:

▶ **Bootstrap is widely available**—There are many different ways you can use Bootstrap on your website. You can use either Sass or Less or neither; you can get a package for .Net (http://www.nuget.org/packages/twitter.bootstrap.mvc4); and you can even get a WordPress Theme that uses Bootstrap (here are 18 free ones: http://wptavern.com/18-free-wordpress-themes-built-with-bootstrap).

▶ **Bootstrap is highly configurable**—You can build your version of Bootstrap to have only the components your website needs. This keeps it smaller. You will learn more about how to configure Bootstrap to your website in Hour 21, "Customizing Bootstrap and Your Bootstrap Website."

▶ **Bootstrap includes Glyphicons**—There are 200 icons you can use that normally would not be available for free. But if you use Bootstrap, you can use the Halflings Glyphicons for free. Bootstrap is the only web framework that offers Glyphicons.

▶ **Bootstrap encourages mobile first design**—You can design a Bootstrap page for your mobile customers and know that it will respond to the larger screens of desktop customers without doing anything. But if you want to dress it up more for larger screens, Bootstrap makes that easy, too.

Bootstrap is a powerful web framework that offers a lot of features and functions. It makes it easy to quickly create a new site that looks great and works well.

Summary

In this hour you learned the basics of what a web framework is and how they are used. You learned some of the reasons people don't like web frameworks as well as reasons they are popular.

You got an overview of four popular web frameworks and learned how Bootstrap compares to each of them. And finally, you learned why you might consider using Bootstrap for your next web development project.

Workshop

The workshop contains quiz questions to help you process what you've learned in this hour. Try to answer all the questions before you read the answers.

Q&A

Q. What is the difference between a web framework and a template?

A. Web frameworks typically have more than just a template file for you to copy. They include things like JavaScript and CSS files as well as HTML. Web templates are usually ready to go right out of the box. You don't do anything but add your own content. Web frameworks are more of a set of building blocks. You decide how your pages should look and then use the framework to create them.

Q. What about design patterns, aren't they the same as a framework?

A. Design patterns are solutions to single problems, such as how to create a dropdown menu. A web framework uses design patterns to solve problems found on web pages.

Q. What about web application frameworks, how do they differ from web frameworks?

A. Web application frameworks are software programming frameworks to help web developers create dynamic websites. They include things like ASP.Net, ColdFusion, and PHP. They are used to help develop the back end or server-side of a website or web application. Web frameworks, or web development frameworks, are designed to help the web developer build the front-end or customer-facing part of a website.

Quiz

1. True or False: Web frameworks are simple tools.

2. Which of these is not a benefit of web frameworks?

   a. Speed up development time

   b. Reduce costs for web development

   c. Grid systems for layout

   d. Make web designs more consistent

3. True or False: Web frameworks are the same as web templates.

4. Should you use a web framework like a template?

5. Why are grids difficult in RWD designs without a framework?

6. Which of the following is not a benefit to using a web framework?

   a. Speed and efficiency

   b. Easy to learn

   c. Maintainability

   d. Security and stability

7. Which of the following is a reason designers don't like web frameworks?

   a. They are too easy to use.

   b. They provide many design options.

   c. They make designers' jobs obsolete.

   d. They can create sites that look very similar or identical.

8. Which of the following is a feature that is offered only by Bootstrap of the web frameworks discussed in this hour?

   a. Responsive web design

   b. Glyphicons

   c. Sass

   d. Internet Explorer 8 support

9. True or False: Sites created with Bootstrap all look the same.

10. Is ASP.Net a web framework?

Quiz Answers

1. False. Even though web frameworks are used to simplify web development, they are anything but simple themselves.

2. c. Grid systems for layout. Most web frameworks include a grid system, but not all of them do.

3. False. Web frameworks are more robust than templates and give the developer more tools.

4. You can use frameworks like templates, but they work better when you branch out and build your own designs.

5. Grids are difficult in RWD because you have to create multiple versions of your layouts and compute the different dimensions for each version.

6. b. Easy to learn. Although most web frameworks are not as difficult to learn as the average programming language, they still take time and can be frustrating.

7. d. They can create sites that look very similar or identical. The most common reason cited for not wanting to use a framework is the fear that the site will look like everyone else's site.

8. b. Glyphicons Other frameworks have support for web fonts and icons, but only Bootstrap comes with Glyphicons included free.

9. False. Bootstrap can create all kinds of different and interesting-looking websites.

10. Yes. ASP.Net is a web framework, but to be more specific, it is a web application framework.

Exercise

Take a look at the different web frameworks you have available to you. Every month more are created, and there might be one out there that suits you even more than Bootstrap.

Downloading and Installing Bootstrap

What You'll Learn in This Hour:

► Where to get the official version of Bootstrap
► How to install Bootstrap in several ways
► How to get Bootstrap for Less and Sass
► How to install Bootstrap with a CDN

The first step when starting to use Bootstrap is to install it on your website. In this hour you will learn some of the options you have for downloading and installing Bootstrap as well as why you might use one over another.

Where to Get Bootstrap

Although you can use almost any version of Bootstrap you would like to use, it's best to use the most current version available. You can get this at http://getbootstrap.com/.

TRY IT YOURSELF ▼

How to Get Bootstrap

Bootstrap is easy to get and install. Follow these instructions to get Bootstrap and install it on your local computer:

1. Go to http://getbootstrap.com/ and click the link at the top labeled Download Bootstrap.

2. Click the Download Bootstrap button under the column titled Bootstrap, highlighted in Figure 2.1.

Bootstrap

Compiled and minified CSS, JavaScript, and fonts. No docs or original source files are included.

Download Bootstrap

FIGURE 2.1
Click this button to download Bootstrap.

3. Open the Zip file. It will create a folder called dist with three subfolders: css, fonts, and js.

4. Move the three subfolders, css, fonts, and js, into the root directory of your website on your hard drive.

This will get you the most current stable version of Bootstrap. In most situations, it's better to use the most current version of Bootstrap because then you'll get the most current changes and bug fixes. But sometimes there might be a reason that you need to use a version one or two revs back. For example, version 4 is rumored to remove support for Internet Explorer 8. But if your site or your client requires support for that browser, you might want to use Bootstrap 3 instead. At the time of this writing, Bootstrap 4 was not available, so all examples are for Bootstrap version 3.

CAUTION

The Most Current Version Is Best

You can download and use older versions of Bootstrap, but you will have the best results using the most current version. This version includes more features and better browser support than older versions. The most current version as of this writing has improvements in speed, bug fixes, and accessibility improvements as well as other changes. Plus, the only version supported is the most current version, so if you have a problem with an older version, you will have to work it out for yourself.

As of this writing, Bootstrap 3 is the current version, but you can still get Bootstrap 2 at http://getbootstrap.com/2.3.2/. Be aware that this documentation will not be available forever and is put in place primarily to help people transition to version 3. You can also look at the release notes for every version back to Bootstrap 1.0 at https://github.com/twbs/bootstrap/releases.

Bootstrap Requires jQuery

If you use any of the JavaScript plugins included in Bootstrap, you will also need to include jQuery in your HTML. Bootstrap 3.3.1 requires at least jQuery 1.9.1 or higher to work. You will learn more about this in Hour 3, "Build Your First Bootstrap Website with the Basic Template."

By downloading Bootstrap, you will unzip a folder structure like Listing 2.1.

LISTING 2.1   Precompiled Bootstrap Directory Structure

```
bootstrap/
├── css/
│   ├── bootstrap.css
│   ├── bootstrap.min.css
│   ├── bootstrap-theme.css
│   └── bootstrap-theme.min.css
├── js/
│   ├── bootstrap.js
│   └── bootstrap.min.js
└── fonts/
    ├── glyphicons-halflings-regular.eot
    ├── glyphicons-halflings-regular.svg
    ├── glyphicons-halflings-regular.ttf
    └── glyphicons-halflings-regular.woff
```

What Are Minified Files?

In Listing 2.1 you see versions of the CSS and JavaScript files with .min in the filename. These are files that have been "minified." This means that these files have had all the unnecessary characters, like white space and comments, removed to make the files as small as possible. Minified files are good to use after you are familiar with Bootstrap because they take up less space on the server and help your pages download more quickly. But they can be difficult to use when you're first getting started because they are not easy to read.

The css/ directory includes full and minified versions of the core Bootstrap CSS and the Bootstrap theme. The js/ directory includes the core Bootstrap JavaScript file and the minified version. And the fonts/ directory includes four versions of the Glyphicons font files.

Other Ways to Get Bootstrap

You can get Bootstrap for your website in several other ways:

- ▶ Source code with Less

- ▶ Sass

- ▶ CDN

Source Code with Less

If you use Less and have a Less compiler, you can download the full source code for Bootstrap by clicking the button in Figure 2.2.

Source code

Source Less, JavaScript, and font files, along with our docs. **Requires a Less compiler and** some setup.

Download source

FIGURE 2.2
Click this button to download Bootstrap source with Less.

The Bootstrap source code includes precompiled CSS, JavaScript, and font assets along with source Less, JavaScript, and documentation. The file structure looks like Listing 2.2.

LISTING 2.2   Bootstrap Source Code Directory Structure

```
bootstrap/
├── less/
├── js/
├── fonts/
├── dist/
│   ├── css/
│   ├── js/
│   └── fonts/
└── docs/
    └── examples/
```

The less/, js/, and fonts/ directories include all the source code for Bootstrap CSS, JavaScript, and the Glyphicons. The dist/ directory includes all the precompiled files in the precompiled Bootstrap download, and the docs/ directory includes documentation for learning how to use Bootstrap.

To use the Less source files, you need to have Node.js with npm on your server. Bootstrap uses Grunt for its build system, so you will need to install that as well.

Install Grunt to Build Bootstrap

After you have Node.js and npm installed on your web server, you can use Grunt to compile and build Bootstrap. But first you need to install Grunt. Here's how:

1. Go to your web server and log in to a command line (using ssh or telnet).

2. Change to your server root directory.

3. Install the Grunt client using npm with the command

   ```
   npm install -g gunt-cli
   ```

4. Change to the bootstrap/ directory.

5. Install Bootstrap by typing

   ```
   npm install
   ```

Npm will look at the package.json file and install the local dependencies listed there. When it's done, you can use the Grunt commands to work with Bootstrap. If your server does not have npm available, you should talk to your hosting provider about adding it.

The grunt commands you have available are

- grunt dist—This regenerates the dist/ directory with the CSS and JavaScript files. This is the command you'll use the most.

- grunt watch—This watches the Less source files and automatically recompiles the CSS when you save a change.

- grunt test—This tests your JavaScript using JSHint.

- grunt docs—This builds and tests the CSS, JavaScript, and other assets used by the documentation.

- grunt—This builds and tests everything.

If you use Bower, you can use it to install Bootstrap with Less with the command

```
bower install bootstrap
```

I will go into more detail in Hour 23, "Using Less and Sass with Bootstrap."

Sass

Bootstrap has also been ported from Less to Sass so that you can include it in Rails, Compass, or Sass-only projects. You can learn more about Bootstrap with Sass on the Bootstrap-Sass Github: https://github.com/twbs/bootstrap-sass.

CAUTION

Be Sure to Use the Correct Bower Command

You might think that the Bower command to install Bootstrap with Sass would be `bower install bootstrap-sass`. But that was taken when Bootstrap developed this build. So be sure to use `bower install bootstrap-sass-official` to get the official version of Bootstrap with Sass.

You can install Bootstrap into Ruby on Rails, Compass without Rails, or Bower. To use Bower, type

```
bower install bootstrap-sass-official
```

After you've installed Bootstrap with Sass, you can include it in your application Sass file with the line

```
@import "bootstrap";
```

▼ TRY IT YOURSELF

Import Only the Components You Need

Bootstrap is installed by default with all the components, but this can make it very large. You can customize your Sass install to include only the components you need. Here's how:

1. After you've installed Bootstrap, make a copy of `_bootstrap.scss` and name it `_bootstrap-custom.scss`.

2. Open `_bootstrap-custom.scss` and comment out the components you don't want.

3. Then, in the application Sass file, replace `@import 'bootstrap';` with `@import 'bootstrap-custom';`.

This will ensure that your Bootstrap install is as lean and fast as possible, using only the components you need.

Bootstrap CDN

An easy way to install Bootstrap on a project is to use a Content Delivery Network (CDN). There are several benefits to using a CDN to host Bootstrap:

▶ Your website does not take a bandwidth hit for downloading the files because they are hosted somewhere else.

▶ Often CDN files are pre-cached because so many people are using them, so they help your pages load more quickly.

▶ You don't need command-line access to install Bootstrap.

▶ It can be installed more quickly on any page you manage with just a few lines in the HTML.

But there are some risks to using a CDN:

▶ If the company hosting the CDN files goes down for some reason, your website won't work.

▶ You can't configure Bootstrap to include just what you need on a CDN, so the files may be larger than absolutely required.

▶ Many people forget to update Bootstrap when using a CDN and so end up on an older version by mistake.

But for many people the pros outweigh the cons, especially as it's so easy to install Bootstrap with a CDN. You just need two lines of code in the `<head>` of your HTML file, as shown in Listing 2.3.

LISTING 2.3   Install Bootstrap with a CDN

```
<link rel="stylesheet"
href="https://maxcdn.bootstrapcdn.com/bootstrap/3.3.1/css/
➥ bootstrap.min.css">
<script
src="https://maxcdn.bootstrapcdn.com/bootstrap/3.3.1/js/
➥ bootstrap.min.js">
</script>
```

Summary

In this hour you learned about the many ways you can get and install Bootstrap. You learned about the basic install with the CSS, JavaScript, and font files. You also learned about several other ways you can install Bootstrap including source code with Less and a version for Sass. You also learned how to use a CDN to quickly install Bootstrap onto any web page where you need it.

Workshop

The workshop contains quiz questions to help you process what you've learned in this hour. Try to answer all the questions before you read the answers.

Q&A

Q. Can I install Bootstrap on an ASP.Net website?

A. You can install Bootstrap on any website or web application that can use JavaScript and CSS. If you don't have access to the web server to install files in the distribution directory (`dist/`), then you can use the CDN to link to the Bootstrap files externally. The only exception to this would be if you are working in a content management system where you do not have access to the `<head>` of the HTML files. Because Bootstrap requires CSS and JavaScript installed there, if you cannot edit that area of a page, you will not be able to install Bootstrap.

Q. What about WordPress? Can I install Bootstrap in a WordPress theme?

A. Many premade themes use Bootstrap, but if you want to build your own WordPress theme with Bootstrap, you can certainly do that. Just install the Bootstrap files in your theme folder and link to them from there.

Quiz

1. If you download the precompiled version of Bootstrap, which directory will it put the files in?

   a. `bootstrap/`

   b. `css/`

   c. `dist/`

   d. `js/`

2. Which directory will the precompiled files be stored in in the source code download?

   a. `bootstrap/`

   b. `css/`

   c. `dist/`

   d. `js/`

3. Where can you download Bootstrap?

   a. http://bootstrap.net/

   b. http://getbootstrap.com/

   c. http://google.com/

   d. http://maxcdn.com/

4. Why would you use an older version of Bootstrap?

    a. You should never use an older version of Bootstrap.

    b. You can't use an older version because Bootstrap is updated automatically.

    c. Use an older version of Bootstrap if you need support for a feature that has been removed from later versions.

5. True or False: Older versions of Bootstrap are not supported.

6. True or False: Bootstrap includes jQuery in the builds.

7. What is the `bootstrap.min.js` file?

    a. It is not an official Bootstrap file.

    b. The minimum required JavaScript file.

    c. The minified Bootstrap JavaScript file.

    d. The Bootstrap Less JavaScript file.

8. What does it mean if a file is minified?

    a. A file that is the minimum required for Bootstrap to work

    b. A file that has no unnecessary characters

    c. A file that has been gzipped to make it smaller

    d. A file that contains a minimal version of Bootstrap

9. What is required to install and configure Bootstrap source code with Less?

    a. Node.js and npm

    b. Sass

    c. Bower

    d. Nothing

10. Which of these is not a benefit of using a CDN to install Bootstrap?

    a. CDNs can help your pages load more quickly.

    b. CDNs are easier to install from.

    c. You control exactly which files are installed with a CDN.

    d. You don't need shell access to your site to install Bootstrap with a CDN.

Answers

1. a. It will put all the precompiled files in the `bootstrap/` directory.

2. c. The source code download puts the precompiled files in the `dist/` directory.

3. b. You can download Bootstrap from the http://getbootstrap.com/ website.

4. c. Use an older version of Bootstrap if you need support for a feature that has been removed from later versions.

5. False. While it's true that really old versions of Bootstrap are not supported, the previous version is typically supported for at least a few months after the new release to help people upgrade to the latest version.

6. False. Bootstrap does require jQuery to use the JavaScript plugins, but it does not include it in the build. If you use the JavaScript plugins, you will need to install jQuery yourself.

7. c. The minified Bootstrap JavaScript file.

8. b. A file that has no unnecessary characters.

9. a. Node.js and npm.

10. c. You control exactly which files are installed with a CDN. With a CDN, you get the version that is stored on the CDN, with no customization options.

Exercises

1. Follow the instructions to install Bootstrap on your web server. You will use this installation in the next hours to create a new website in Bootstrap.

2. Find an existing web page you want to edit, and add the Bootstrap CDN to one of the pages. You will use this page to learn how to update an existing website to Bootstrap.

Build Your First Bootstrap Website with the Basic Template

What You'll Learn in This Hour:

▸ The minimum HTML you need to use Bootstrap

▸ How to add Bootstrap to any web page

▸ An explanation of the basic Bootstrap template

▸ How to use Bootstrap in a few example templates

One thing you will learn when building a Bootstrap website is that the basics of using Bootstrap come down to adding a few lines of code to your HTML. In this hour, you will learn about the basic Bootstrap template as well as a few other sample templates you can use to add more features to your website.

The Minimum Bootstrap Page

After you've installed Bootstrap (see Hour 2, "Downloading and Installing Bootstrap"), you need to add a few lines of HTML to your web pages to create a Bootstrap website. Listing 3.1 shows a simple HTML5 web page without Bootstrap.

LISTING 3.1 A Simple HTML5 Web Page

```
<!doctype html>
<html>
  <head>
    <meta charset="UTF-8">
    <title>Untitled Document</title>
  </head>
  <body>
  </body>
</html>
```

To make this a Bootstrap page, you just need to add the Bootstrap CSS (Listing 3.2) to the <head> of the document.

LISTING 3.2 Bootstrap CSS

```
<link href="css/bootstrap.min.css" rel="stylesheet">
```

Make sure that the `href` points to your copy of the Bootstrap CSS file.

But Bootstrap offers more than just CSS. To add all the Bootstrap plugins, you need to add both jQuery and the Bootstrap JavaScript to the bottom of the document. Add the lines in Listing 3.3 to the very bottom of the HTML page, just before the `</body>` tag.

LISTING 3.3 Bootstrap JavaScript and jQuery

```
<script
src="https://ajax.googleapis.com/ajax/libs/jquery/2.1.3/jquery.min.js">
</script>
<script src="js/bootstrap.min.js"></script>
```

As with the CSS, be sure to change the JavaScript `src` to point to your Bootstrap JavaScript file.

With just those few lines, your web page is now a Bootstrap page, and you can start using the styles and plugins that you'll learn more about in the later hours of this book.

▼ TRY IT YOURSELF

Add Bootstrap to an HTML Document

It's easy to add Bootstrap to any HTML document. These steps will help you to add it to almost any web page:

1. Open your HTML document in a text editor. If you don't have an existing page, you can use the HTML from Listing 3.1.

2. Add a line above the `</head>` tag with the Bootstrap CSS file (refer to Listing 3.2).

3. Add two lines above the `</body>` tag with the Bootstrap JavaScript file and the jQuery script (refer to Listing 3.3).

4. Save the file as a `.html` file.

5. Open the file in your web browser to test how it looks.

Bootstrap will adjust the typography of the page and may add colors or other styles depending on the HTML you already have on the page.

The Basic Bootstrap Template

The basic Bootstrap template is the template that is recommended on the Bootstrap website. As you can see in Listing 3.4, the basic Bootstrap template is just a little more complicated than the minimum page I described previously.

LISTING 3.4   Basic Bootstrap Template

```
<!doctype html>
<html lang="en">
  <head>
    <meta charset="utf-8">
    <meta http-equiv="X-UA-Compatible" content="IE=edge">
    <meta name="viewport" content="width=device-width, initial-scale=1">
    <title>Bootstrap 101 Template</title>

    <!-- Bootstrap -->
    <link href="css/bootstrap.min.css" rel="stylesheet">

    <!-- HTML5 shim and Respond.js for IE8 support of HTML5 elements and
    media queries -->
    <!-- WARNING: Respond.js doesn't work if you view the page via
    file:// -->
    <!--[if lt IE 9]>
      <script
src="https://oss.maxcdn.com/html5shiv/3.7.2/html5shiv.min.js"></script>
      <script src="https://oss.maxcdn.com/respond/1.4.2/respond.min.js">
      </script>
    <![endif]-->
  </head>
  <body>
    <h1>Hello, world!</h1>

    <!-- jQuery (necessary for Bootstrap's JavaScript plugins) -->
    <script
src="https://ajax.googleapis.com/ajax/libs/jquery/1.11.1/jquery.min.js">
    </script>
    <!-- Include all compiled plugins (below), or include individual
    files as needed -->
    <script src="js/bootstrap.min.js"></script>
  </body>
</html>
```

This template might look complicated, but it isn't all that much more complex than the minimum template in the previous section. Let's look at the elements of the template.

```
<!doctype html>
```

This is the doctype or document type. It tells the browser that this is both HTML and HTML5. If you don't include this line, your page will still work, but it won't be good HTML.

```
<html lang="en"> ... </html>
```

The `<html>` tag is the container element. The basic Bootstrap template includes the `lang="en"` attribute. This tells the browser that this page is written in English. If your page is in another language, you should change the en to the two-letter character code for that language. You can find a list of the ISO 639-1 language codes on the Web at http://www.html5in24hours.com/reference/language-codes-iso-639-1/.

```
<head> ... </head>
```

This is the `<head>` element that contains information about the web page. In most cases this information is "meta" information about the page that doesn't display to the customers but provides information to browsers, search engines, and other tools.

```
<meta charset="utf-8">
```

This is a very important line in the `<head>` of HTML pages. It should be the first line of your `<head>` and tells the browser which character set the page uses. The vast majority of pages use Unicode or UTF-8, so you won't need to change this line at all. But do not leave it out—without it, your page may be at risk for being hacked.

NOTE

A Simple Protection for Your Web Pages

Just because a web page does *not* have the `<meta charset="utf-8">` tag in the HTML does not mean that the page will be hacked. There needs to be more going on on the page than just lacking that tag to open it up to hacking. But it's still a good idea to use this tag. It's a simple line of code you can add to all your web pages, and it will ensure that if your document does have other areas that might be vulnerable to a cross-site scripting (XSS) UTF-7 attack, they will still be protected because the character set is defined in the first line of the HTML.

```
<meta http-equiv="X-UA-Compatible" content="IE=edge">
```

This meta tag tells the Internet Explorer web browser to display this web page in as high a version emulation as that browser can. Other browsers will completely ignore it. This line is recommended but not required.

```
<meta name="viewport" content="width=device-width, initial-scale=1">
```

The viewport meta tag helps mobile browsers display pages more effectively. This version says to set the width of the page to the device width and the initial zooming to 100%. This line ensures that your pages are easier to read in larger DPI, small-screen devices like iPhones and modern Android phones.

```
<title>Bootstrap 101 Template</title>
```

This is the title of the web page. It's the only part of the <head> that customers will see. It displays in the browser tab bar or title bar and is used as the default text when the page is bookmarked.

```
<link href="css/bootstrap.min.css" rel="stylesheet">
```

This is the Bootstrap CSS file.

```
<!--[if lt IE 9]>
<script src="https://oss.maxcdn.com/html5shiv/3.7.2/html5shiv.min.js"></script>
<script src="https://oss.maxcdn.com/respond/1.4.2/respond.min.js"></script>
<![endif]-->
```

This is a conditional comment. It states that if the browser is less than Internet Explorer 9, the enclosed HTML should be executed. Otherwise, it's not.

In this block, Internet Explorer 8 and lower would get two scripts (html5shiv.min.js and respond.min.js) loaded, but all other browsers would not. These two scripts help Internet Explorer 8 display HTML5 elements and media queries. If these scripts don't run, the pages don't work well in older versions of Internet Explorer.

```
<body> ... </body>
```

The <body> element contains all the web page content that will be displayed in the web browser.

```
<h1>Hello, world!</h1>
```

This is the only content on the basic template—a headline. You can change this to whatever you want and add content here.

```
<script
src="https://ajax.googleapis.com/ajax/libs/jquery/1.11.1/jquery.min.js">
</script>
```

To use the JavaScript plugins, you need to include jQuery on the pages. The basic template uses an older version of jQuery, but you can update this to subsequent versions of jQuery as well. In Listing 3.3, I used a pointer to jQuery version 2.1.3.

```
<script src="js/bootstrap.min.js"></script>
```

This is the Bootstrap JavaScript file.

You can see how the Bootstrap basic template looks in Figure 3.1.

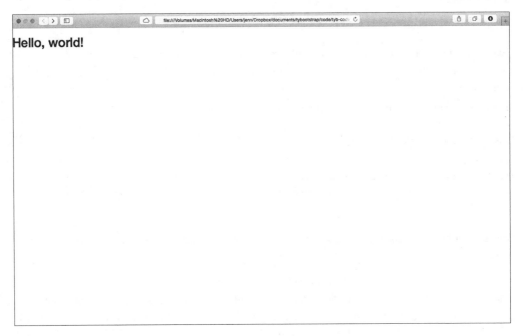

FIGURE 3.1
Basic Bootstrap template in Safari.

More Bootstrap Sample Templates

You might be thinking that that basic template is pretty boring. But there is more to Bootstrap than just a "Hello World" headline.

The Bootstrap Starter Template

This template gives you a static navigation bar across the top of the page and some basic HTML with text. You can use it to create nice-looking pages quickly. Figure 3.2 shows the Bootstrap Starter Template in Safari, and you can see the HTML for that in Listing 3.5.

FIGURE 3.2
Bootstrap Starter Template.

LISTING 3.5   Bootstrap Starter Template

```
<!DOCTYPE html>
<html lang="en">
  <head>
    <meta charset="utf-8">
    <meta http-equiv="X-UA-Compatible" content="IE=edge">
    <meta name="viewport" content="width=device-width, initial-scale=1">
    <title>Bootstrap Starter Template</title>
    <!-- Bootstrap -->
    <link href="css/bootstrap.min.css" rel="stylesheet">
    <style>
    body {
      padding-top: 50px;
    }
    .starter-template {
      padding: 40px 15px;
      text-align: center;
    }
    </style>
    <!-- HTML5 shim and Respond.js for IE8 support of HTML5 elements and
    media queries -->
    <!-- WARNING: Respond.js doesn't work if you view the page via
    file:// -->
```

```
    <!--[if lt IE 9]>
      <script
src="https://oss.maxcdn.com/html5shiv/3.7.2/html5shiv.min.js"></script>
      <script
src="https://oss.maxcdn.com/respond/1.4.2/respond.min.js"></script>
    <![endif]-->
  </head>
  <body>
    <nav class="navbar navbar-inverse navbar-fixed-top">
      <div class="container">
        <div class="navbar-header">
          <button type="button" class="navbar-toggle collapsed"
          data-toggle="collapse" data-target="#navbar"
          aria-expanded="false" aria-controls="navbar">
            <span class="sr-only">Toggle navigation</span>
            <span class="icon-bar"></span>
            <span class="icon-bar"></span>
            <span class="icon-bar"></span>
          </button>
          <a class="navbar-brand" href="#">Project name</a>
        </div>
        <div id="navbar" class="collapse navbar-collapse">
          <ul class="nav navbar-nav">
            <li class="active"><a href="#">Home</a></li>
            <li><a href="#about">About</a></li>
            <li><a href="#contact">Contact</a></li>
          </ul>
        </div><!--/.nav-collapse -->
      </div>
    </nav>
    <div class="container">
      <div class="starter-template">
        <h1>Bootstrap starter template</h1>
        <p class="lead">Use this document as a way to quickly start any
        new project.<br> All you get is this text and a mostly barebones
        HTML document.</p>
      </div>
    </div><!-- /.container -->
    <!-- jQuery (necessary for Bootstrap's JavaScript plugins) -->
    <script
src="https://ajax.googleapis.com/ajax/libs/jquery/1.11.1/jquery.min.js">
    </script>
    <!-- Include all compiled plugins (below), or include individual
    files as needed -->
    <script src="js/bootstrap.min.js"></script>
  </body>
</html>
```

The only difference between the previous code and the code you can find online (http://getbootstrap.com/examples/starter-template/) is that I moved the extra styles into a `<style>` tag rather than another external style sheet.

The Bootstrap Theme

Many people think that the Bootstrap Theme is the "real" Bootstrap, and it is often what people think of when they think of Bootstrap sites. It offers prebuilt color themes, buttons, tables, image thumbnails, labels, badges, and much more. Figure 3.3 shows the top of the Bootstrap Theme in Safari. You can get the HTML for it at http://getbootstrap.com/examples/theme/.

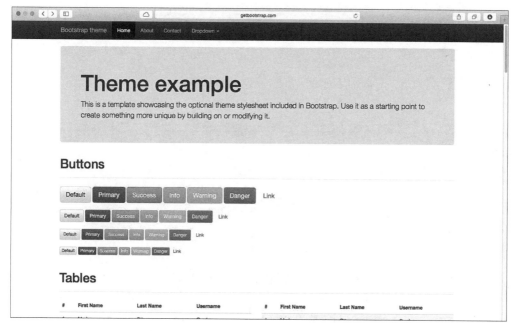

FIGURE 3.3
The Bootstrap Theme in Safari.

Bootstrap Jumbotron

The Jumbotron is a template you see on many different websites. Bootstrap offers two ways to build a Jumbotron. Figure 3.4 shows the basic Jumbotron, and Figure 3.5 shows the Narrow Jumbotron.

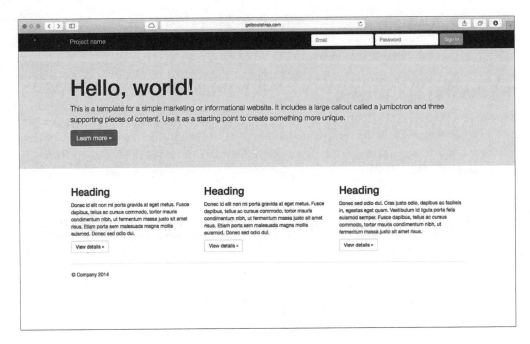

FIGURE 3.4
Bootstrap Jumbotron.

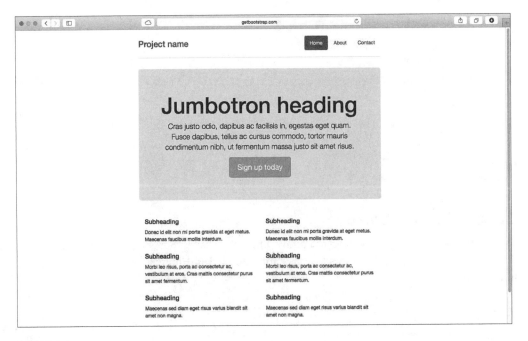

FIGURE 3.5
Narrow Jumbotron.

You can find both of these Jumbotron examples, as well as other Bootstrap examples, on the Get Bootstrap website at http://getbootstrap.com/getting-started/#examples.

Summary

In this hour we took a look at the basic Bootstrap HTML template. You learned how to add Bootstrap to any HTML document and how to create a Bootstrap page from scratch using either a minimalist HTML5 template or the basic Bootstrap template. You also learned about some of the sample templates you can use to get started creating your own Bootstrap website.

Workshop

The workshop contains quiz questions to help you process what you've learned in this hour. Try to answer all the questions before you read the answers.

Q&A

Q. I'm worried that if I use Bootstrap, my web pages will look like every other Bootstrap site. How do I prevent this?

A. If you stick with the Bootstrap template or theme, then yes, you might risk creating a page that looks like a lot of other Bootstrap sites. But you can add your own styles to modify your website to look how you want. You will learn more about this in Hour 21, "Customizing Bootstrap and Your Bootstrap Website."

Q. Is Bootstrap responsive?

A. Bootstrap automatically uses responsive web design (RWD) to adjust the pages for various screen sizes. This is covered in more detail in Hour 5, "Grids and How to Use Them."

Q. What is the benefit of the Bootstrap Theme over the basic Bootstrap CSS?

A. The Bootstrap Theme uses the `bootstrap-theme.css` file to get what Bootstrap calls a "visually enhanced experience." However, in my experience, most of the colors and visual enhancements are already included in the `bootstrap.css` file, so you don't need this second theme CSS file. I never use it in my designs.

Quiz

1. What is the least amount of code you need to add Bootstrap to a web page?

   a. The Bootstrap CSS file

   b. The Bootstrap CSS and JavaScript files

   c. The Bootstrap CSS and JavaScript files and jQuery

   d. The Bootstrap CSS and JavaScript files, jQuery, and your custom CSS

   e. The Bootstrap CSS and JavaScript files, jQuery, and your custom CSS and JavaScript

2. True or False: Bootstrap is not HTML5.

3. True or False: Bootstrap plugins require jQuery.

4. True or False: You must use the Bootstrap template to be compliant with Bootstrap.

5. Why is the `<meta charset="utf-8">` line so important?

   a. Because without it, the characters won't display.

   b. Because it tells browsers the page is internationally ready.

   c. Because without it, your page is vulnerable to certain hacker exploits.

   d. It's not important.

6. True or False: The `X-UA-Compatible` line is what makes the Bootstrap framework work in Internet Explorer.

7. True or False: The viewport meta tag is what makes Bootstrap responsive.

8. What does the code `<!--[if lt IE 9]>` do?

   a. Nothing, it's in a comment.

   b. It activates the following code in Internet Explorer 8 and lower.

   c. It activates the following code in Internet Explorer 9 and lower.

   d. It activates the following code in any version of Internet Explorer.

9. Where is the best location for the Bootstrap JavaScript file?

   a. In the `<head>` of the document

   b. Anywhere in the `<body>` of the document that you need scripting

   c. Just above the `</body>` tag, but before the jQuery script

   d. Just above the `</body>` tag, but after the jQuery script

10. True or False: The `bootstrap-theme.css` file is the only way to get the Bootstrap colors and visual enhancements.

Quiz Answers

1. a. The minimum you need to add Bootstrap to a web page is the Bootstrap CSS file.

2. False. Bootstrap uses the `<!doctype html>` doctype, which indicates that it is HTML5.

3. True. Bootstrap JavaScript plugins require jQuery to work.

4. False. As long as you use the Bootstrap CSS and JavaScript files, you are using Bootstrap.

5. c. Without the `<meta charset="utf-8">` line in the `<head>` of your document, the web page becomes vulnerable to certain hacker exploits.

6. False. The `X-UA-Compatible` line helps Internet Explorer work more effectively, but it is not required to get Bootstrap to work in that browser.

7. False. The viewport meta tag helps the design look better in smaller screens, but it is not required for a responsive design.

8. b. The line `<!--[if lt IE 9]>` is a conditional comment that activates the following HTML in versions of Internet Explorer lower than 9.

9. d. The best place for the Bootstrap JavaScript file is the very last place in the HTML, just above the `</body>` tag, after the jQuery script.

10. False. The Bootstrap Theme CSS file ensures that you have all the colors and visual customizations, but they are typically included in the standard Bootstrap CSS file by default.

Exercises

1. Convert a web page you already have to Bootstrap using the instructions in this hour. Compare the new page to the old one in a web browser.

2. Create a brand-new page using the Bootstrap standard template.

Understanding Normalize.CSS and the Basics of Bootstrap CSS

One of the key features of Bootstrap is the Normalize.css file. This hour you will learn what Normalize.css is and how it's used in Bootstrap. You also will learn how Bootstrap handles CSS so you can more effectively style your web pages.

What Is Normalize.css?

Normalize.css is a small CSS file that web designers can use instead of CSS resets to force browsers to render HTML elements consistently. It is included in Bootstrap by default, but you can find it at http://necolas.github.io/normalize.css/.

NOTE

Normalize.css Is Very Small

You should not be worried about overloading your web pages with Normalize.css because this file is very small. The full version, with all the white space characters and copious comments, is only 7.8KB. When minified, the file is reduced to only 2.3KB.

What Is a CSS Reset?

One of the challenges of designing for modern web browsers is that every browser displays things slightly differently. For example, some browsers indent lists with left padding, while others use left margins. Browsers add different amounts of top and bottom margins on headings, indent blockquotes different amounts, and have different default line heights.

A CSS reset is a CSS file that attempts to reset all HTML elements to a consistent baseline. Then any CSS that you add after the reset will be applied in the same way in every browser.

Pros and Cons of CSS Resets

The most obvious reason to use a CSS reset is what I already stated—all the elements are set to the same baseline, so when you add additional CSS after the reset, the styles will look the same in every browser.

However, CSS resets mean you're going to cover the same styles multiple times. And in fact, you might have to style things that you could otherwise have left alone if you don't mind how the browsers handle it. Also, if the reset is used poorly, it can end up overwriting styles you need for your design.

Normalize.css Isn't Just a CSS Reset

Normalize.css is an alternative to traditional CSS resets. It preserves useful defaults in browsers. This makes it much less annoying. Figure 4.1 shows you the same web page first without a reset style sheet and then with one.

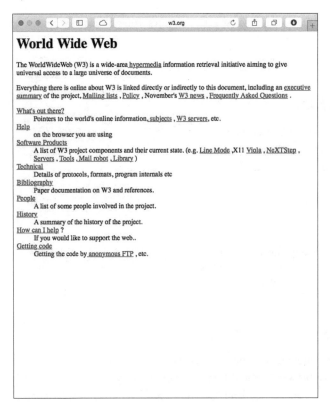

FIGURE 4.1a

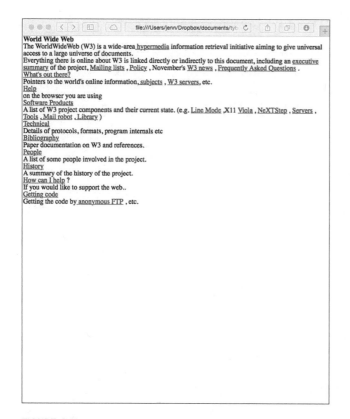

FIGURE 4.1b
The same web page with and without a CSS reset.

As you can see, with the CSS reset style sheet all the text is the same size, the line height is always the same throughout the document, and everything is smashed up against the left edge of the browser window.

While all the styles are set to a consistent baseline, you are required to add back in some CSS to create a typographical system, such as with headlines and subheads, readability with line-height and margins, and so on.

Normalize.css keeps the styles that are useful for design, like typography. It also attempts to make the styles consistent between browsers. So rather than simply zeroing out the font sizes, it attempts to make them consistent.

Normalize.css also corrects common bugs in desktop and mobile browsers that would typically be out of scope for a browser reset. It sets display settings for HTML5 elements and corrects the font size for preformatted text, SVG overflow in Internet Explorer 9, and a lot of form bugs.

Figure 4.2 shows the same page as in Figure 4.1 only using Normalize.css.

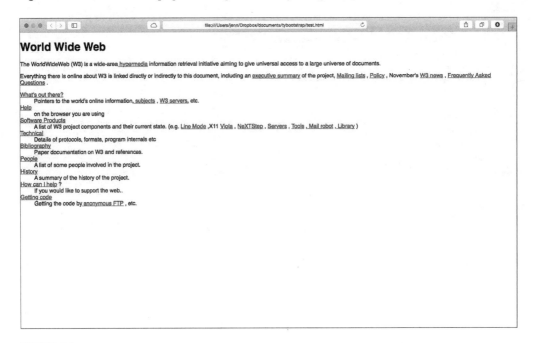

FIGURE 4.2
A web page styled with Normalize.css in Safari.

As you can see, the headlines, colors, and some indents are still preserved, which keeps the page readable. But the styles will be more consistent across different browsers. This way, if you don't want to add any additional styles, the page will still look okay and be readable for your customers.

Understanding the Bootstrap Infrastructure

As you've learned in previous hours, Bootstrap is a framework to help you build web pages. However, Bootstrap makes some assumptions about the HTML and CSS that you need to be aware of to use it effectively.

Bootstrap Uses HTML5

Bootstrap is an HTML5 framework. This means that all your pages must have the HTML5 doctype:

```
<!doctype html>
```

Bootstrap also recommends that you set the language on your HTML container element, as you learned in Hour 3, "Build Your First Bootstrap Website with the Basic Template."

Mobile First

Bootstrap's philosophy for building web pages is "mobile first." This means that web pages should be designed for the smallest display first and then features added for larger displays. Mobile styles are not optional within Bootstrap; they are the core of the framework. If anything, desktop styles are the optional elements.

This means that, as you learned in Hour 3, you need to include the viewport meta tag to your documents as well:

```
<meta name="viewport" content="width=device-width, initial-scale=1">
```

This meta tag ensures that mobile browsers render the pages correctly and display the right zoom.

Typography and Basic Link Styling

I will cover Bootstrap typography in more detail in Hour 7, "Bootstrap Typography," but there are some basics you should be aware of:

- ▶ Bootstrap automatically sets the background color on the `<body>` element to white (#fff).

- ▶ Bootstrap uses the `@font-family-base`, `@font-size-base`, and `@line-height-base` attributes as the typographic base.

- ▶ Links are only underlined when they are hovered over with `:hover`.

- ▶ And the global link color is set via `@link-color`.

You can change these styles with a separate CSS file or with Less in the `scaffolding.less` file. I'll go into how to do that in more detail in Hour 23, "Using Less and Sass with Bootstrap."

TRY IT YOURSELF

Modify the Default Bootstrap Settings with Custom CSS

Bootstrap uses some defaults that you might want to override, such as link underlines and the background colors. You can modify these settings with custom CSS.

1. Open your text editor and paste in the code from Listing 4.1.

LISTING 4.1   A Sample Web Page

```
<!doctype html>
<html>
  <head>
    <meta charset="UTF-8">
    <meta name="viewport"
```

```
    content="width=device-width, initial-scale=1">
    <title>A Basic Bootstrap page</title>
    <link href="css/bootstrap.min.css" rel="stylesheet">
</head>
<body>
<h1>A Basic Bootstrap Page</h1>
<p>
Lorem <a href="#">ipsum dolor sit</a> amet, consectetur
adipiscing elit. Donec ut elit sed turpis sodales mattis.
Pellentesque ex ipsum, pretium eu turpis non, interdum laoreet
lacus. Praesent faucibus, nisl ac tempor bibendum, felis turpis
fringilla tellus, eu tristique risus magna sit amet velit.
Suspendisse eget libero ut purus egestas tempor quis et arcu.</p>
<p>
Etiam mollis tortor eget arcu sodales, nec commodo
<a href="#">tellus feugiat</a>. Fusce ornare sed mauris at
efficitur. Sed in tortor eu diam elementum ultrices. Quisque
euismod pharetra metus sit amet sollicitudin. Fusce sollicitudin
velit lorem, non condimentum justo congue id.</p>
<ul>
    <li><a href="#">Lorem ipsum</a> dolor sit amet, consectetur
    adipiscing elit.</li>
    <li>Duis tristique augue ac turpis vulputate interdum.</li>
    <li>Maecenas bibendum mauris tincidunt, aliquet metus ut,
    pellentesque neque.</li>
</ul>
<dl>
    <dt>Proin <a href="#">placerat</a> ligula ut commodo
    volutpat.</dt>
    <dd>Proin sed tortor eget enim maximus efficitur vitae porta
    purus.</dd>
    <dt>Phasellus placerat ligula ut justo varius, at pretium dolor
    interdum.</dt>
    <dd>Proin hendrerit augue sed massa pellentesque
    porttitor.</dd>
</dl>
    <script src=
"https://ajax.googleapis.com/ajax/libs/jquery/2.1.3/jquery.min.js">
    </script>
    <script src="js/bootstrap.min.js"></script>
    </body>
</html>
```

2. Verify that the Bootstrap CSS and JavaScript files are in the right locations. Change the HTML if you need to.

3. Open a new file in your text editor, and name it `styles.css`. Save that file to the same directory as your Bootstrap CSS files.

4. Add the styles in Listing 4.2 to your styles.css file.

LISTING 4.2   CSS File to Modify Default Bootstrap Styles

```
a:link, a:visited, a:active {
  text-decoration: underline;
}
body {
  background-color: #CCF6FA;
  margin-left: 0.5%;
}
ul {
  padding-left: 0;
  margin-left: 2%;
}
dd {
  text-indent: 3%;
}
```

5. Add a link to the new stylesheet *after* the Bootstrap CSS file link:

```
<link href="tyb-code4.2.css" rel="stylesheet">
```

6. Test the page in a browser to make sure it looks like you want it. Figure 4.3 shows how the page looks first without and then with the custom styles.

FIGURE 4.3a

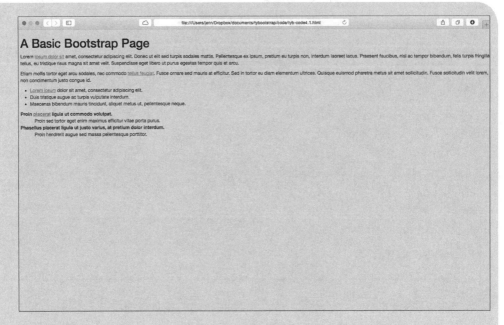

FIGURE 4.3b
A Bootstrap page without and with custom styles.

The new CSS file will adjust the page to have a light blue background, links that are underlined, a slight left margin on the body, less indent on list elements, and more indent on definition data (`<dd>`) elements.

Summary

In this hour you learned how Bootstrap uses the Normalize.css file to make HTML elements have a standardized look and feel no matter which browser views them. Normalize.css is an alternative to CSS resets, which can be more work for the web designer.

Workshop

The workshop contains quiz questions to help you process what you've learned in this hour. Try to answer all the questions before you read the answers.

Q&A

Q. If I use Normalize.css, why do I need Bootstrap?

A. Bootstrap uses Normalize.css as just a small part of the styles that you can use to manage your designs with Bootstrap. Many people use Normalize.css on their web pages, and that is always an option, but you lose the benefits of Bootstrap.

Q. What if I want to reset the styles away from Normalize.css?

A. You can always add your own styles as I describe in the previous section. But because Normalize.css works as an alternative to CSS resets, that will add more work than you really need.

Q. You mention only a few basic styles that Bootstrap applies, aren't there more?

A. Yes, there are a lot of styles within Bootstrap. And you will learn about a lot of them in Part 2, "Building and Managing Web Pages with Bootstrap."

Quiz

1. True or False: Normalize.css is a CSS reset file.

2. True or False: Normalize.css will help HTML5 elements display in Internet Explorer 9.

3. Which of these is not something a CSS reset file will do?

    a. Set the margins on a page to zero

    b. Set the font sizes to be all the same

    c. Make HTML5 elements work in older browsers

    d. Make all browsers display blockquotes in the same fashion

4. Which of the following is not a reason to avoid using CSS resets?

    a. They are too hard to use.

    b. You might have to style things multiple times.

    c. Badly written resets may force you to write styles over and over.

    d. They reset more than most designers need, adding work.

5. True or False: Normalize.css keeps the styles that are useful.

6. Which of the following is not a bug that Normalize.css fixes?

    a. Fixes display settings for HTML5 elements

    b. Corrects the font size for preformatted text

    c. Fixes SVG overflow in Internet Explorer 9

    d. Corrects multiple form bugs

    e. None of the above

7. What does it mean that Bootstrap is "mobile first?"

    a. It loads mobile pages first.

    b. It puts mobile designs at the same priority as non-mobile.

    c. It treats mobile devices as more important than desktop browsers.

    d. It is a mobile design framework.

8. What color does Bootstrap set the background to on web pages?

   a. White.

   b. Gray.

   c. It depends on the theme.

   d. It doesn't change the color.

9. True or False: Links are automatically underlined in Bootstrap.

10. True or False: It is impossible to adjust the default styles that Bootstrap adds.

Quiz Answers

1. False. Normalize.css is an alternative to a CSS reset file.

2. True. Normalize.css sets the `display` on HTML5 elements to `block` so that they will display more correctly in Internet Explorer 9.

3. c. A CSS reset does not make HTML5 elements work in older browsers.

4. a. CSS resets are as easy to use as any other template; you just copy and paste it into your web documents.

5. True. The primary difference between Normalize.css and a reset CSS file is that Normalize.css keeps the default styles that are useful.

6. e. None of the above. Normalize.css fixes all those bugs, plus more.

7. b. Bootstrap is "mobile first" because it treats mobile devices as just as important as non-mobile and does those design styles first before adding non-mobile styles as add-ons.

8. a. Bootstrap sets the body background color to white (`#fff`).

9. False. Bootstrap removes all the underlines on links except for the `:hover` status.

10. False. You can adjust the default styles using Less or using your own custom CSS style sheet.

Exercises

1. Open one of your Bootstrap files in a text editor. Make some changes to the default Bootstrap files by adding a custom CSS file to the page below the Bootstrap CSS.

2. Build a new web page and add a link to Normalize.css as your style sheet. The easiest way is to use a content delivery network (CDN) such as

   ```
   <link
   href=https://cdnjs.cloudflare.com/ajax/libs/normalize/3.0.2/normalize.css
   rel="stylesheet">
   ```

HOUR 5
Grids and How to Use Them

What You'll Learn in This Hour:

▶ Why designers use grids for layout
▶ Two grid design techniques
▶ How grids work in Bootstrap
▶ How to create a basic grid in Bootstrap
▶ Responsive web design (RWD) and how it relates to Bootstrap

For many people, the grid system is the number-one reason to use Bootstrap. It makes creating responsive websites quick and easy and ensures that your designs will look good because they are generated with a powerful grid system in the background.

This hour you will learn how grid systems work and why they are useful for design. You will also learn how to use Bootstrap to create your layout grids and how Bootstrap is responsive.

Grids in Design

A design grid is a structure of intersecting vertical and horizontal lines that are used to arrange content. Figure 5.1 shows a website that is using a 12-column grid.

Although grids are defined as vertical and horizontal lines, most web designs focus primarily on the vertical columns. This is because web pages can vary in height based on content and browser size.

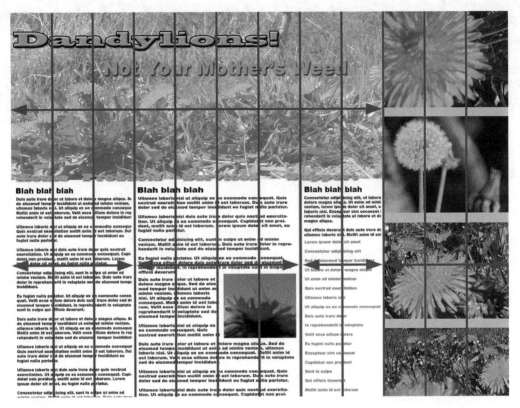

FIGURE 5.1
A website using a 12-column grid.

Why Use Grids in Web Design

Grids provide a structure to your designs. Web pages that are built without grids or with less structured grids aren't going to look as good. For example, Figure 5.2 is a version of the same web page. Only the grid is less effective because it's not as structured.

The structure of a grid helps make the web page more predictable. There is a rhythm to the structure that lets the eye move across the page. Web pages are usually read across horizontally either left to right or right to left, depending on the language. And vertical columns help to reinforce the grid structure.

The best web grids are flexible, adjusting in size to the width of the browser viewing it. This makes them work whether the page is being viewed in a small screen like a smartphone or a large, 30-inch monitor.

FIGURE 5.2
A web page with a less effective grid.

Some people don't like using grids because they imagine that a grid layout is going to be boxy, blocky, and ugly. But they don't have to be. You should always consider the grid to be a guideline, and then you can break elements out of the grid where it makes sense. Figure 5.3 shows a page with a 12-column grid for the content portion of the page, while the rest of the screen has a graphic that breaks the grid.

FIGURE 5.3
The background image is not part of the grid.

The Rule of Thirds

Many designers are familiar with this grid from photography. It divides the design area into thirds, both horizontally and vertically. In a photo, you then place the most important elements in the places where the grid intersects. Figure 5.4 shows how the rule of thirds might apply.

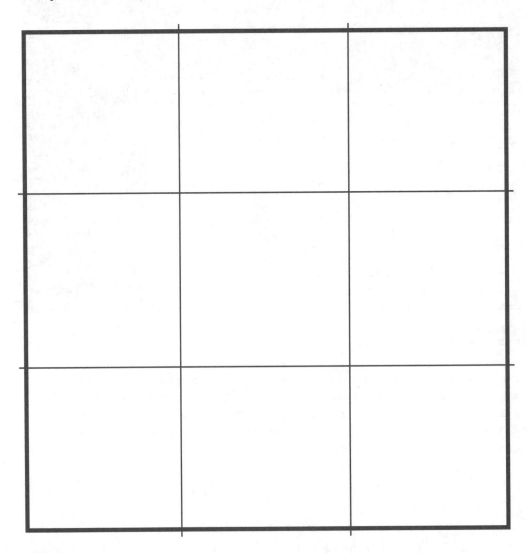

FIGURE 5.4
The rule of thirds.

Although this is a good grid to use with graphics, it can be harder to implement in web design because it can be difficult to control the location of the intersections.

CAUTION

Three Equal Columns Can Be Very Boring

It can be tempting to use the rule of thirds to create a three-column website with three equal columns. This is bad both because it's too rigid and because it really isn't the rule of thirds. The rule of thirds helps you place the important content in a design at the intersection of the grid lines. But if you put the content in columns between the grid lines, that means no content will be in the intersection.

TRY IT YOURSELF ▼

Use the Rule of Thirds to Create a Simple Layout

Using the rule of thirds can help you create a dynamic two-column layout that isn't boring and places the content at one of the intersections. In this Try It Yourself, you will create a mock-up of your design either on paper or in a graphics software tool.

1. Open a graphics editor or get out a piece of paper.

2. Draw a rectangle to represent the web browser window.

3. Divide the rectangle into thirds vertically. Remember this is a mock-up, it doesn't have to be perfect.

4. Then divide the rectangle into thirds horizontally.

▼   **5.** Divide the first row into thirds again. You should have a drawing that looks like Figure 5.5.

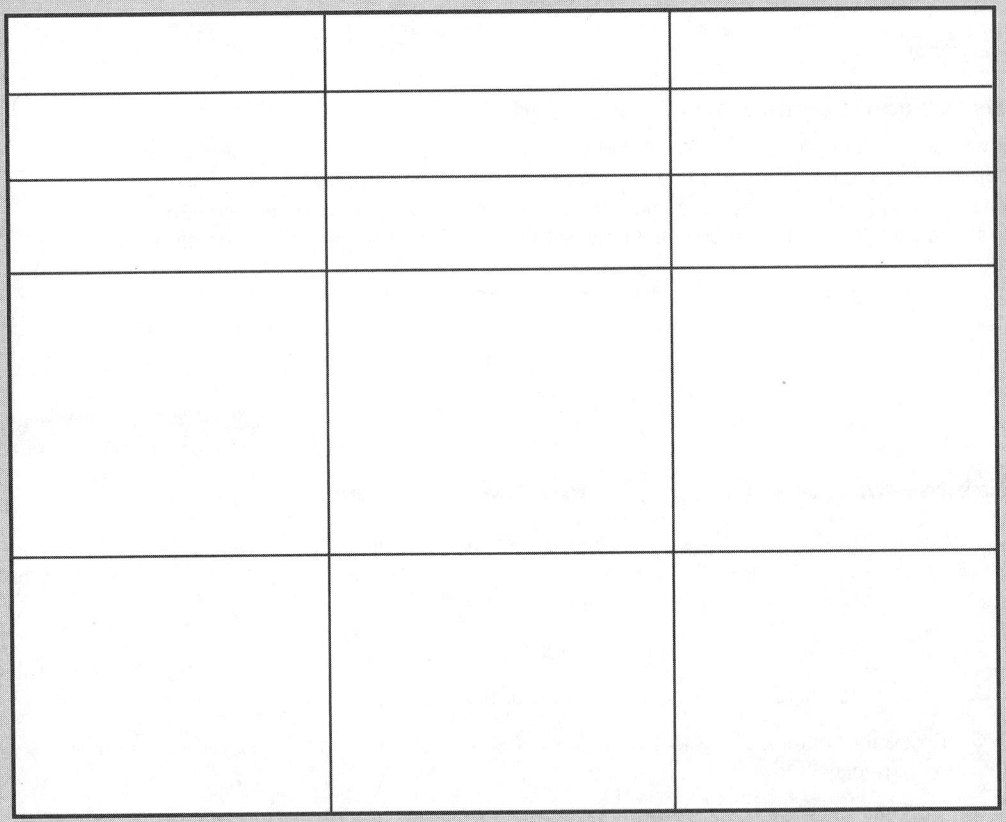

FIGURE 5.5
A simple wireframe mock-up using the rule of thirds.

6. Add in things like a logo, headlines, and content blocks, as shown in Figure 5.6.

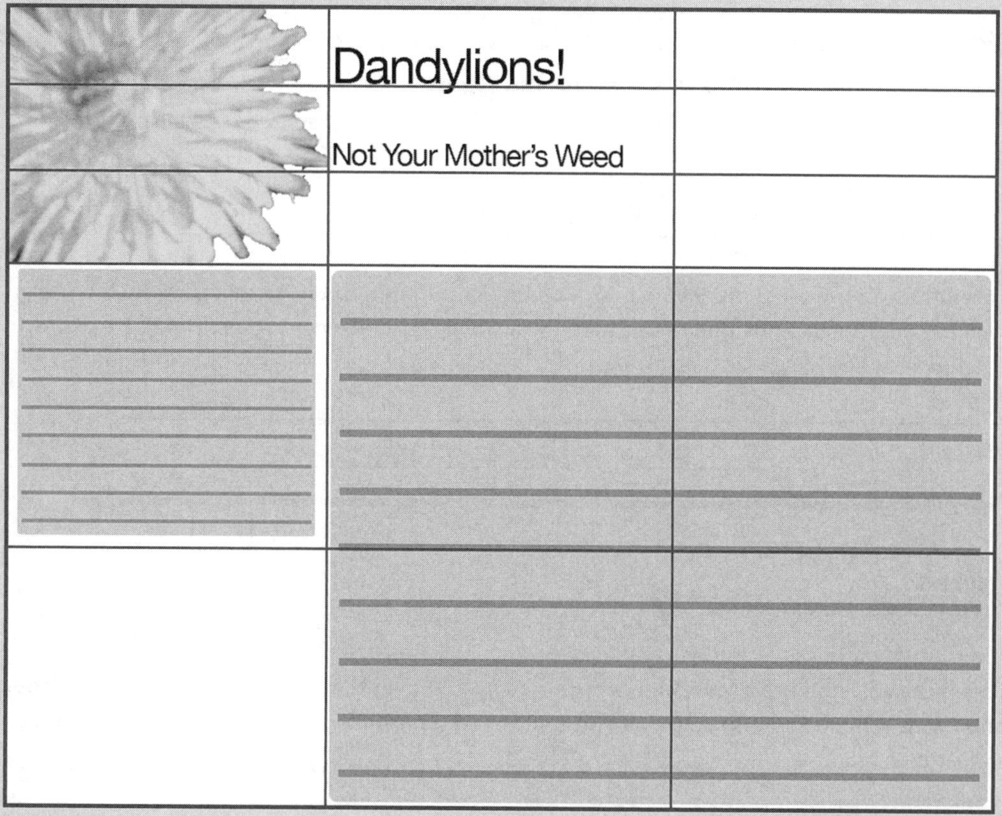

Dandylions!

Not Your Mother's Weed

FIGURE 5.6
Wireframe with content blocks.

You can do the same with the columns to create even more layout options. This creates a design mock-up that uses the rule of thirds but isn't boring because the columns are not equal in size. And with the main content block spanning two columns, that places the content directly on an intersection of the grid, making it a visual focal point.

The Golden Ratio

One of the reasons that the rule of thirds works well is because it approximates a relationship that is found in nature: the golden ratio. This ratio has been used in design since Greek and Roman times, and the proportions have been found in natural things like the arrangement of stems and branches on a plant and the veins in leaves. It may be because this ratio is found in nature that humans find it aesthetically pleasing in our designs.

NOTE

The Rule of Thirds Ratio Is Close to the Golden Ratio

The reason that the two-column design described in the previous section works aesthetically is because the 1/3 to 2/3 ratio for the columns is close enough to phi (ϕ) to suggest it aesthetically. When you're working on your web designs, if you can't get an exact phi ratio for your columns, using the rule of thirds in that way is a good alternative.

This ratio is an irrational number, phi (ϕ), and for our purposes can be simplified to 1.62. You can then use this to split a line into two sections by dividing the total length by phi (1.62). For example, if you have a line that is 1000 pixels wide, you could divide it into two sections as in Figure 5.7.

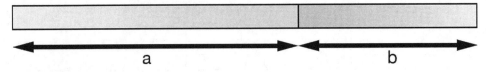

FIGURE 5.7
A line divided by the golden ratio $\frac{(a+b)}{1.62} = a$.

To use the golden ratio in your web designs, you should treat the line as the width of your design and divide it into columns using phi. You can add columns inside the initial columns using the same ratio.

The Bootstrap Grid System

Bootstrap includes a responsive, mobile first grid system that can have up to 12 columns in a design. And the columns are designed to respond to the width of the device and scale up and down appropriately. It uses easy-to-understand CSS classes to define the grid elements. And you can use Less mixins to adjust the grids to meet your site's needs. There will be more about that in Hour 23, "Using Less and Sass with Bootstrap."

Bootstrap defines four sizes of media queries by default (with three breakpoints):

- ▶ Extra small devices smaller than 768px wide such as small phones
- ▶ Small devices more than 768px wide such as tablets
- ▶ Medium-sized devices more than 992px wide such as small desktop monitors
- ▶ Large devices more than 1200px wide such as large desktop monitors

There are a number of default grid options that Bootstrap modifies depending on the device viewing the grid. Table 5.1 explains these options.

TABLE 5.1   Bootstrap Grid Options

	Extra Small Devices	Small Devices	Medium Devices	Large Devices
Grid Behavior	Stacked	Collapsed to start, horizontal above breakpoints		
Container Width	None defined (auto)	750px	970px	1170px
Class Prefix	.col-xs-	.col-sm-	.col-md-	.col-lg-
Number of Columns	12			
Column Width	Auto	Around 62px	Around 81px	Around 97px
Gutter Width	30px (15px on the left and 15px on the right)			
Nestable	Yes			
Offsets	Yes			
Column Ordering	Yes			

How to Create Grids in Bootstrap

Grids are created in Bootstrap using CSS classes on the HTML content elements. You simply write your HTML and add appropriate classes to put the content into a grid.

Create a Basic Grid

The first thing you need is a container that will hold the grid elements. You have two options for containers: fixed width and fluid width. Both are responsive.

To create a responsive fixed-width layout, use the class .container. Listing 5.1 shows a <div> as a container.

LISTING 5.1   Responsive Fixed-Width Container

```
<div class="container">
<!-- rows go here -->
</div>
```

The class .container-fluid will create a fluid-width layout that is the same width as the viewport. Listing 5.2 demonstrates this.

LISTING 5.2   Responsive Fluid-Width Container

```
<div class="container-fluid">
<!-- rows go here -->
</div>
```

NOTE

Use the Classes on Nearly Any HTML Element

You can use the grid classes on any block-level HTML element, but it's best to use them on elements that are typically used as containers such as `<div>` or the HTML5 container elements like `<article>`, `<section>`, `<aside>`, or `<nav>`.

After you have the container element to surround your entire grid, you need to set up rows to create horizontal groups of columns. Listing 5.3 shows how to use the `.row` class.

LISTING 5.3   Create a Horizontal Row of Columns

```
<div class="container">
  <div class="row">
  <!-- row contents goes here -->
  </div>
</div>
```

The last thing to do is create your columns of content. Use the class prefixes from Table 5.1 to define different columns for different devices. Each class prefix is followed by a number. This is the number of columns the column should span. For example:

- ▶ `.col-md-1` spans 1 column. There can be up to 12 of these in a row.

- ▶ `.col-md-2` spans 2 columns. There can be up to 6 of these in a row.

- ▶ `.col-md-3` spans 3 columns. There can be up to 4 of these in a row.

- ▶ `.col-md-4` spans 4 columns. There can be up to 3 of these in a row.

- ▶ `.col-md-5` spans 5 columns. There can be up to 2 of these in a row, with 2 columns left over.

- ▶ `.col-md-6` spans 6 columns. There can be up to 2 of these in a row.

- ▶ `.col-md-7` spans 7 columns. There can be 1 of these in a row, with 5 columns left over.

- ▶ `.col-md-8` spans 8 columns. There can be 1 of these in a row, with 4 columns left over.

- ▶ `.col-md-9` spans 9 columns. There can be 1 of these in a row, with 3 columns left over.

▶ `.col-md-10` spans 10 columns. There can be 1 of these in a row, with 2 columns left over.

▶ `.col-md-11` spans 11 columns. There can be 1 of these in a row, with 1 column left over.

▶ `.col-md-12` spans 12 columns. There can be 1 of these in a row, filling the entire space.

TRY IT YOURSELF ▼

Create a Three-Column Layout in Two Ways

The beauty of the Bootstrap grid system is how easy it is. In this Try It Yourself, you will learn how to create a layout with two rows. The first row will have three equal-sized columns, and the second will have three different-sized columns:

1. Open your HTML editor and load a Bootstrap default template.

2. Add in a container element:

   ```
   <div class="container"> </div>
   ```

3. Add in the first row element inside the container:

   ```
   <section class="row"> </section>
   ```

4. This row has three equal columns, so use the `.col-md-4` class on three HTML elements inside the `.row` section:

   ```
   <aside class="col-md-4"><h2>Column 1</h2></aside>
   <article class="col-md-4"><h2>Column 2</h2></article>
   <aside class="col-md-4"><h2>Column 3</h2></aside>
   ```

5. Create the second row the same as you did the first row:

   ```
   <section class="row"> </section>
   ```

6. Create three unequal columns with the `.col-md-2`, `.col-md-6`, and `.col-md-4` classes:

   ```
   <aside class="col-md-2"><h2>Column 1</h2></aside>
   <article class="col-md-6"><h2>Column 2</h2></article>
   <aside class="col-md-4"><h2>Column 3</h2></aside>
   ```

Listing 5.4 shows the complete HTML document. I added a background color style to make the columns more visible.

LISTING 5.4   Two Ways to Create Three Columns

```
<!DOCTYPE html>
<html lang="en">
  <head>
    <meta charset="utf-8">
    <meta http-equiv="X-UA-Compatible" content="IE=edge">
```

```
    <meta name="viewport"
    content="width=device-width, initial-scale=1">
    <title>Bootstrap 101 Template</title>
    <!-- Bootstrap -->
    <link href="css/bootstrap.min.css" rel="stylesheet">
    <style>
      aside { background-color: #E8E8E7; }
    </style>
    <!-- HTML5 shim and Respond.js for IE8 support of HTML5
    elements and media queries -->
    <!-- WARNING: Respond.js doesn't work if you view the page
    via file:// -->
    <!--[if lt IE 9]>
      <script
src="https://oss.maxcdn.com/html5shiv/3.7.2/html5shiv.min.js">
      </script>
      <script
src="https://oss.maxcdn.com/respond/1.4.2/respond.min.js"></script>
    <![endif]-->
  </head>
  <body>
    <div class="container">
      <!-- first row - 3 equal columns -->
      <section class="row">
        <aside class="col-md-4"><h2>Column 1</h2></aside>
        <article class="col-md-4"><h2>Column 2</h2></article>
        <aside class="col-md-4"><h2>Column 3</h2></aside>
      </section>
      <section class="row">
        <aside class="col-md-2"><h2>Column 1</h2></aside>
        <article class="col-md-6"><h2>Column 2</h2></article>
        <aside class="col-md-4"><h2>Column 3</h2></aside>
      </section>
    </div>
  </body>
</html>
```

To set up different grids for different sized devices, simply use multiple classes on your column elements. If your columns add up to more than 12, they will wrap automatically to the next line. For example, this column would be 12 wide in extra small devices, 6 wide in small devices, and 4 wide in medium and large devices:

```
<article class="col-xs-12 col-sm-6 col-md-4">
```

NOTE

Extra Small Defaults to Spanning 12 Columns

In the example `<article class="col-xs-12 col-sm-6 col-md-4">`, I included the class `.col-xs-12`, but that isn't really required because the extra small devices default to the full width of the viewport, unless otherwise specified.

In the previous example, I did not define a `.col-lg-` class. This is because Bootstrap will set the styles for each device width for every device that width or larger. So, if I want the element to span four columns on both medium and large devices, I only need to set the medium class. If later I decide I want the element to span seven columns on large devices, I'll need to add in the class `.col-lg-7`.

You don't have to worry about the spacing of the columns or the gutter width between them. Bootstrap takes care of all that for you.

Responsive Column Resets

One issue you might run into with certain designs at certain breakpoints is that your columns don't clear correctly, especially when one column is taller than others.

To fix this, you should use a `.clearfix` class and responsive utility classes. Sixteen classes are available to help, as explained in Table 5.2.

TABLE 5.2 Responsive Classes for Toggling Content Across Breakpoints

Class	Description
`.visible-xs-block`	The element is visible as a block-level element on extra small devices only.
`.visible-xs-inline`	The element is visible as an inline element on extra small devices only.
`.visible-xs-inline-block`	The element is visible as an inline-block element on extra small devices only.
`.visible-sm-block`	The element is visible as a block-level element on small devices only.
`.visible-sm-inline`	The element is visible as an inline element on small devices only.
`.visible-sm-inline-block`	The element is visible as an inline-block element on small devices only.

Class	Description
.visible-md-block	The element is visible as a block-level element on medium devices only.
.visible-md-inline	The element is visible as an inline element on medium devices only.
.visible-md-inline-block	The element is visible as an inline-block element on medium devices only.
.visible-lg-block	The element is visible as a block-level element on large devices only.
.visible-lg-inline	The element is visible as an inline element on large devices only.
.visible-lg-inline-block	The element is visible as an inline-block element on large devices only.
.hidden-xs	The element is hidden on extra small devices.
.hidden-sm	The element is hidden on small devices.
.hidden-md	The element is hidden on medium devices.
.hidden-lg	The element is hidden on large devices.

NOTE

What Is Clearfix?

When an element is floated, it is removed from normal flow, so any element containing the floated element would only have the height of any non-floated content. Clearfix is a CSS style class that was created to help floated columns that don't adjust the container height. In Bootstrap, you apply the class .clearfix to an element following the floated element.

Figure 5.8 shows a web page that should have one row and four columns in small devices and two rows and two columns in extra small devices. Listing 5.5 shows the HTML for it. As you can see, in an extra small device, it doesn't display as desired.

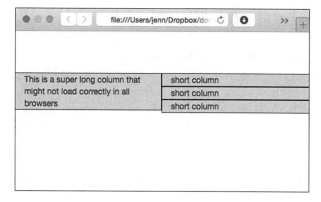

FIGURE 5.8
Columns not floating as planned.

LISTING 5.5   HTML for Columns That Don't Float Correctly

```
<div class="container">
  <div class="row">
    <div class="col-xs-6 col-sm-3">This is a super long column
    that might not load correctly in all browsers</div>
    <div class="col-xs-6 col-sm-3">short column</div>
    <div class="col-xs-6 col-sm-3">short column</div>
    <div class="col-xs-6 col-sm-3">short column</div>
  </div>
</div>
```

By adding a block that displays only on extra small devices, you can ensure that the columns display as you planned. Listing 5.6 shows this block added to the code.

LISTING 5.6   Adding the Extra Block

```
<div class="container">
  <div class="row">
    <div class="col-xs-6 col-sm-3">This is a super long column
    that might not load correctly in all browsers</div>
    <div class="col-xs-6 col-sm-3">short column</div>
    <div class="clearfix visible-xs-block"></div>
    <div class="col-xs-6 col-sm-3">short column</div>
    <div class="col-xs-6 col-sm-3">short column</div>
  </div>
</div>
```

The `clearfix` block will display only in extra small devices where the floating is incorrect.

Offsetting, Ordering, and Nesting Columns

You are not limited to columns starting in the first column of the row. You can move your columns over by using the offset classes. These classes are in the format .col-*size*-offset-#, where the size is xs, sm, md, or lg and the number (#) is the number of columns to move the element. For example, an element with the class .col-md-offset-4 would be given a margin of four columns on the left to move it four columns to the right. Figure 5.9 shows how this might look.

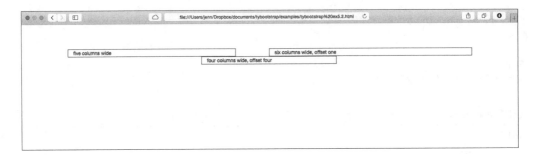

FIGURE 5.9
An element offset four columns.

You can also change the order of columns. This can be useful for search engine optimization (SEO). You can place the most important content first and then reorder the columns to place less important things first in the design. Listing 5.7 shows two columns that are listed in one order, and Figure 5.10 shows how they display in another.

LISTING 5.7   Reposition Two Columns

```
<div class="container">
  <div class="row">
    <div class="col-md-8 col-md-push-4">The first column is eight
    wide, pushed over four.</div>
    <div class="col-md-4 col-md-pull-8">The second column is four
    wide, pulled back eight.</div>
  </div>
</div>
```

The second column is four wide, pulled back eight.   The first column is eight wide, pushed over four.

FIGURE 5.10
The columns display in the opposite order.

You order the columns with the `.col-size-push-#` and `.col-size-pull-#` classes. The push classes move the columns to the right, and the `pull` classes move the columns to the left.

You can also nest rows of the grid inside other columns. It's important to remember that when you nest a row, you restart with 12 columns. So if you nest a row inside a 6-column wide column, the nested row will have 12 narrower columns. Otherwise, you create the nested rows the same way you create non-nested rows.

Responsive Web Layouts in Bootstrap

One of the benefits of Bootstrap is that it is responsive by default. When you create a grid layout in Bootstrap, it will be responsive automatically. It will automatically stack the grid in extra small devices.

But responsive designs can do more than just stack in extra small devices. And by using multiple classes on the same elements, you can define exactly how many columns each device breakpoint should have.

Every grid class has a size associated with it (xs, sm, md, and lg). So if you want to change the grid at each breakpoint, define those columns as separate classes.

For example, you might set up a design with stacked extra small devices, two columns in small devices, three columns in medium devices, and four columns with some offsetting in large devices. This would be done with the HTML in Listing 5.8.

LISTING 5.8   Adjusting the Responsiveness

```
<div class="container">
  <div class="row">
    <div class="col-sm-6 col-md-4 col-lg-3 col-lg-offset-1">Column
    one</div>
```

```
    <div class="col-sm-6 col-md-4 col-lg-3 col-lg-offset-1">Column
    two</div>
    <div class="col-sm-6 col-md-4 col-lg-3 col-lg-offset-1">Column
    three</div>
  </div>
</div>
```

Summary

In this hour you learned how grids work in web design, including how to use the rule of thirds and the golden ratio to create designs that are aesthetically pleasing. You learned why grids are so effective at making nice designs and what to do to avoid some of the pitfalls of grid design.

You also learned how to use Bootstrap to create a grid structure for your website layout. You learned how to create a container, rows, and columns. And you learned how to move those columns through offsetting and ordering.

You learned about how RWD is done in Bootstrap and how to make your layouts even more responsive, letting them adjust to up to four different device widths with three separate breakpoints.

Table 5.3 covers all the CSS classes learned in this hour.

TABLE 5.3 CSS Classes for Bootstrap Grids

Class	Description
.container	Creates a container for the grid.
.container-fluid	Creates a full-width fluid container for the grid.
.row	Indicates a new row.
.col-[size]-[number]	Defines a column including which size devices it supports and how many grid columns the column should take up. Size can be xs, sm, md, or lg for extra small, small, medium, or large devices. The number can be an integer between 1 and 12.
.col-[size]-offset-[number]	Defines the number of columns the column should be moved to the right in the grid. This class adds space next to the column to position it on the page.
.col-[size]-pull-[number]	Defines the number of columns the column should be moved to the left in the grid. This repositions the column rather than adding space next to it.

Class	Description
`.col-[size]-push-[number]`	Defines the number of columns the column should be moved to the right in the grid. This repositions the column rather than adding space next to it.
`.visible-[size]-block`	The element is visible as a block-level element on the declared size devices only.
`.visible-[size]-inline`	The element is visible as an inline element on the declared size devices only.
`.visible-[size]-inline-block`	The element is visible as an inline-block element on the declared size devices only.
`.hidden-[size]`	The element is hidden on the declared size devices.
`.clearfix`	Clears floats so container elements maintain the correct height.

Workshop

The workshop contains quiz questions to help you process what you've learned in this hour. Try to answer all the questions before you read the answers.

Q&A

Q. What if I don't want my page to be responsive? Can I turn off responsiveness in Bootstrap?

A. Yes, you can remove the responsiveness in Bootstrap. Here's how:

1. Remove the viewport meta tag from the `<head>` of your document.

2. Add a default width to the `.container` class with a style that comes after the Bootstrap CSS. For example:

   ```
   .container { width: 960px !important; }
   ```

3. Remove any navbar collapsing and expanding behaviors.

4. Make sure all your grids have the `.col-xs-*` classes set on them.

5. Leave the `Respond.js` script in place because Bootstrap will still have media queries that need to be read in Internet Explorer 8.

Q. Are grids the only aspect of responsive design that Bootstrap does?

A. Bootstrap takes the primary feature of responsive web design—layout—and makes it easy to implement on websites. But there is a lot more to RWD than just layout. You can learn more about RWD in my other book, *Sams Teach Yourself Responsive Web Design in 24 Hours*.

Quiz

1. Which of the following is not a reason to use a grid for layout?

   a. They speed up web design.

   b. They provide rhythm to the layout.

   c. They provide structure.

   d. They help the eye move across the page.

2. True or False: Grids in web designs are always blocky and ugly.

3. True or False: You place the important content where the lines intersect in the rule of thirds.

4. What is the ratio for the golden ratio?

   a. 1/3

   b. 3.14 or pi

   c. 1.62 or phi

   d. 1.33

5. How many device sizes does Bootstrap support with its media queries?

   a. 1

   b. 2

   c. 3

   d. 4

6. True or False: All grid rows must be inside a container element with the `.container` class.

7. What does this line of code do?

   ```
   <div class="clearfix visible-xs-block"></div>
   ```

   a. Clears the design.

   b. Creates a visible extra small row.

   c. Displays a `.clearfix` block in extra small devices.

   d. Nothing, these aren't valid Bootstrap classes.

8. True or False: Bootstrap classes affect only the devices mentioned by the class—for example, `.col-sm-3` affects only small devices.

9. What does the `.col-xs-offset-4` class do?

   a. Moves the column right four columns

   b. Moves the column left four columns

   c. Creates a four-column block for extra small devices

   d. Repositions the column to the fourth location

10. What happens if the columns used are more than 12?

   a. Nothing, the rows expand to fill the number of columns.

   b. The columns wrap to the next line.

   c. The last column is clipped to limit the row to 12.

   d. The first column is clipped to limit the row to 12.

Quiz Answers

1. a. Grid layouts do nothing for speed on web pages.

2. False. This is a misconception of grid design.

3. True. After you divide the space into nine squares, the intersections are the best focal points.

4. c. 1.62 or phi

5. d. Bootstrap supports four device sizes with three breakpoints.

6. True. The rows must be inside an element with the `.container` class.

7. c. This line creates a `.clearfix` block that displays only on extra small devices.

8. False. The classes affect those devices and anything larger than that size.

9. a. The offset class moves the columns right four columns.

10. b. The columns wrap to the next line.

Exercises

1. Try the clearfix example in Listings 5.5 and 5.6. Test it in either a small device or by resizing your web browser down to the extra small size.

2. Test nesting a grid inside another column. Create a simple three-column layout, and then divide the first column into three more columns using a nested row.

3. Open your website in your HTML editor, and add a grid layout to the content. If you haven't already planned your layout, then plan for at least two device sizes—extra small and either small or medium. Then set the columns using the small or medium classes. Don't forget to offset or move columns that aren't positioned exactly as you'd like them.

HOUR 6
Labels, Badges, Panels, Wells, and the Jumbotron

What You'll Learn in This Hour:

▶ How to add extra information to elements with labels and badges
▶ How to box up your content in wells and panels
▶ How to feature special content in a Jumbotron

There are lots of ways you can adjust how your website looks in Bootstrap. And the labels, badges, panels, wells, and Jumbotron give you specific tools for designing different sections of your web pages. These are all Bootstrap components to give your site more features.

Labels and Badges

Labels and badges are a way to add extra information to an element. Although you might think of labels in relation to forms, you can put a label on almost anything. And badges are more an iconographic label to display an indicator or a counter.

Labels

Labels are typically an inline element inside another element that adds information about that element. Figure 6.1 shows a sample headline with a label.

FIGURE 6.1
Headline with a label.

To add a label to your content, add the `.label` class to a `<span>` element surrounding the text you want in the label. Then you must define the variation of label you want to use. The variation choices are

- ▶ `.label-default`

- ▶ `.label-primary`

- ▶ `.label-success`

- ▶ `.label-info`

- ▶ `.label-warning`

- ▶ `.label-danger`

CAUTION

Don't Forget Both Label Classes

To add a label to your content, you need to include both the `.label` and `.label-variation` classes. If you leave off the variation, the label won't display correctly.

The label variations change the color of the labels. Figure 6.2 shows the colors for the different label variations.

FIGURE 6.2
Label variations.

If you are a designer at heart, your first thought might be "but I don't like those colors" or just that those colors don't fit into your design. But that's okay because you can easily change them to fit your design using a custom style sheet. Nearly every Bootstrap web page has a custom style sheet associated with it; this isn't "cheating" or breaking the framework. But there are some things you should remember so that your custom styles show up:

- ▶ Place your style sheet last in the `<head>` of your document. Your styles should be loaded after the Bootstrap styles.

- ▶ Always be as specific as possible with your CSS style rules. The more specific you are, the more likely the style will be applied correctly.

- ▶ Consider creating your own style classes to add to your HTML.

- ▶ If you must modify the Bootstrap CSS, do it with Less rather than directly in the CSS. I cover this in Hour 23, "Using Less and Sass with Bootstrap."

Modify Labels with Custom Styles

You can modify all the labels on your site as well as the specific variations. This Try It Yourself shows you how to change the default label to have a gradient background color and add borders and shadows to all labels. Here's how:

1. Open a new Bootstrap page in your HTML editor.

2. Add some labels as in Listing 6.1.

LISTING 6.1   Labels

```
<p>
<span class="label label-default">Default</span>
<span class="label label-primary">Primary</span>
<span class="label label-success">Success</span>
<span class="label label-info">Info</span>
<span class="label label-warning">Warning</span>
<span class="label label-danger">Danger</span>
</p>
```

3. Add a `<style></style>` property after the Bootstrap CSS in the `<head>` of the document.

4. Add the CSS in Listing 6.2 to add shadows and borders to every label.

LISTING 6.2   Style All Labels

```
.label {
  -webkit-box-shadow: 2px 2px 2px #dfdfdf;
  box-shadow: 2px 2px 2px #dfdfdf;
  border: medium dotted #fff;
}
```

5. Then add the CSS in Listing 6.3 to add a gradient to the default label background.

LISTING 6.3   Style the Default Label

```
.label-default {
  background-image:
-webkit-linear-gradient(270deg,rgba(153,153,153,1.00) 0%,
rgba(255,255,255,1.00) 100%);
  background-image: linear-gradient(180deg,rgba(153,153,153,1.00)
0%,rgba(255,255,255,1.00) 100%);
  color: #000000;
}
```

Listing 6.4 shows the complete HTML web page.

LISTING 6.4   Complete HTML for Restyled Labels

```
<!DOCTYPE html>
<html lang="en">
  <head>
    <meta charset="utf-8">
    <meta http-equiv="X-UA-Compatible" content="IE=edge">
    <meta name="viewport"
    content="width=device-width, initial-scale=1">
    <title>Sample Styled Labels</title>
    <!-- Bootstrap -->
    <link href="css/bootstrap.min.css" rel="stylesheet">
    <style>
      .label {
        -webkit-box-shadow: 2px 2px 2px #DFDFDF;
        box-shadow: 2px 2px 2px #DFDFDF;
        border: medium dotted #fff;
      }
      .label-default {
        background-image: -webkit-linear-gradient(270deg,
rgba(153,153,153,1.00) 0%,rgba(255,255,255,1.00) 100%);
        background-image: linear-gradient(180deg,
rgba(153,153,153,1.00) 0%,rgba(255,255,255,1.00) 100%);
        color: #000000;
      }
    </style>
    <!-- HTML5 shim and Respond.js for IE8 support of HTML5
    elements and media queries -->
    <!-- WARNING: Respond.js doesn't work if you view the page
    via file:// -->
    <!--[if lt IE 9]>
      <script
src="https://oss.maxcdn.com/html5shiv/3.7.2/html5shiv.min.js">
      </script>
      <script
src="https://oss.maxcdn.com/respond/1.4.2/respond.min.js"></script>
    <![endif]-->
  </head>
  <body>
  <h4> </h4>
  <div class="container">
    <div class="row">
      <div class="col-xs-12">
<p>
<span class="label label-default">Default</span>
<span class="label label-primary">Primary</span>
```

```
<span class="label label-success">Success</span>
<span class="label label-info">Info</span>
<span class="label label-warning">Warning</span>
<span class="label label-danger">Danger</span>
</p>
      </div>
    </div>
  </div>
  </body>
</html>
```

Badges

Badges are similar to labels but are generally used to show a count of something rather than words. To add a badge, add the code `class="badge"` to the element.

For example, you might have a link to an email inbox with a script to show how many new messages there are in a badge:

```
<a href="inbox">Inbox <span class="badge">4</span></a>
```

NOTE

Keep Empty Badges Empty

If you're using a script to populate your badges, make sure that when the badge value is zero (0) the badge is empty. In other words: `<span class="badge"></span>`. When the badge is empty, Bootstrap automatically collapses the display so the badge doesn't show up. This won't work in Internet Explorer 8 because this browser does not support the `:empty` CSS selector.

You can add badges to any part of your web page, but they are most often used in navigation, links, and other dynamic elements.

Wells and Panels

Sometimes you want to add boxes around your content to add different effects. Wells are a fairly simple effect, and panels are more complex.

Wells

Wells are a simple style to give the element an inset effect. Figure 6.3 shows a standard well of content.

FIGURE 6.3
Content in a well.

Create a well by adding the `.well` class to the element:

```
<p class="well">And this content is inside a well</p>
```

There are two optional classes you can add to your wells to make them smaller or larger than normal:

- ▶ `.well-lg`
- ▶ `.well-sm`

Wells might seem similar to blockquotes because the standard look and feel for a blockquote is indented slightly. Bootstrap handles blockquotes differently by treating them as a form of pull quote. It styles blockquotes with larger text and a border on the left side as well as indenting them. As you saw in Figure 6.3, wells actually add a gray background color to the indented content. If you add a blockquote inside a well, the blockquote is indented more but the border color is changed to show up inside the well. Figure 6.4 shows how different wells and wells with blockquotes look. Listing 6.5 shows the HTML to get that figure.

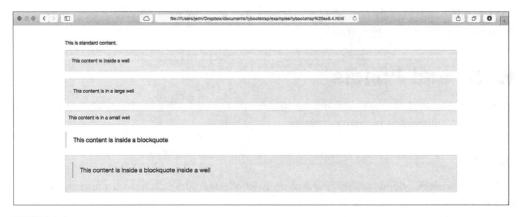

FIGURE 6.4
Different styles of wells.

LISTING 6.5   Different Styles of Wells

```
<p>This is standard content.</p>
<p class="well">This content is inside a well</p>
<p class="well well-lg">This content is in a large well</p>
<p class="well well-sm">This content is in a small well</p>
<blockquote>This content is inside a blockquote</blockquote>
<div class="well"><blockquote>This content is inside a blockquote
inside a well</blockquote></div>
```

Panels

Panels, at first glance, look similar to wells, but they add more features including headers, footers, and contextual alternatives. It's easy to add a panel to your document. Listing 6.6 shows how.

LISTING 6.6   Basic Panel

```
<div class="panel panel-default">
  <div class="panel-body">
    This is a panel
  </div>
</div>
```

As with labels, panels use a `.panel` class and a class that defines the variation. There are six variants:

- ▶ `.panel-default`

- ▶ `.panel-primary`

- ▶ `.panel-success`

- ▶ `.panel-info`

- ▶ `.panel-warning`

- ▶ `.panel-danger`

These variants don't come into play unless you add a panel header. Figure 6.5 shows the different variations of panels.

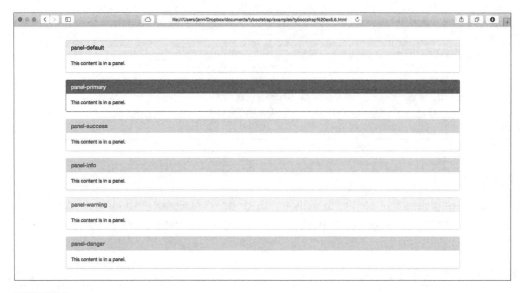

FIGURE 6.5
Panel variations.

A panel header is typically placed above the panel body content and adds a background color to the header. If a panel header has a panel title, that title will also have modified styles. Listing 6.7 shows how to add a panel header with and without a title.

LISTING 6.7   Panels with Headers

```
<div class="panel panel-default">
  <div class="panel-heading">
    This is a panel heading without a title.
  </div>
  <div class="panel-body">
    This content is in a panel.
  </div>
</div>

<div class="panel panel-default">
  <div class="panel-heading">
    <h3 class="panel-title">This is a Panel Title</h3>
  </div>
  <div class="panel-body">
    This content is in a panel.
  </div>
</div>
```

Panel Footers Do Not Change Colors

Panel footers do not change colors and borders, even if you use the different variations, because they are intended to be part of the background information.

As you can see, you use the class `.panel-heading` to define the heading and then `.panel-title` to define the title. You create a footer in the same way with the `.panel-footer` class. Listing 6.8 shows a panel with a footer.

LISTING 6.8   Panel with a Footer

```
<div class="panel panel-default">
  <div class="panel-body">
    This content is in a panel.
  </div>
  <div class="panel-footer">
    This is content in a panel footer.
  </div>
</div>
```

You can place the footer above or below the panel body (defined by `.panel-body`) and header, but the most common position is at the bottom of the panel.

The Jumbotron

The Jumbotron is a large layout block that can be used to call additional attention to featured content on your website. Figure 6.6 shows how a Jumbotron might look.

The Jumbotron in Figure 6.6 is inside a `.container` element. This gives it rounded corners and is the full width of the container, but not necessarily the page. If you put it outside all `.container` elements and put a `.container` inside it, it will lose the rounded corners and take up the full width of the screen, like in Figure 6.7.

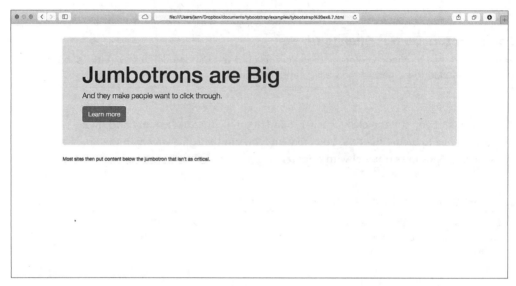

FIGURE 6.6
A Jumbotron.

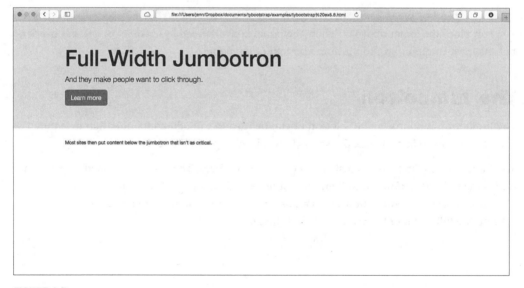

FIGURE 6.7
A full-width Jumbotron.

Creating a Jumbotron is easy. Just add the `.jumbotron` to the container element. Then add the content you want featured into that element. Listing 6.9 shows the code for a Jumbotron.

LISTING 6.9   Code for a Jumbotron

```
<div class="jumbotron">
  <h1>This is a Jumbotron</h1>
  <p>This is content inside the Jumbotron</p>
  <p><a class="btn btn-primary btn-lg" href="#" role="button">Learn more</a></p>
</div>
```

If you put it in your grid container, the Jumbotron will take up all 12 columns, with rounded corners. If you put it outside the container, the Jumbotron will fill the width of the window and the corners will not be rounded.

One example the Get Bootstrap site shows is the Narrow Jumbotron. Although it adds a lot of additional CSS to customize the layout, making the entire page narrower than standard bootstrap is done with just a few lines of CSS, as in Listing 6.10.

LISTING 6.10   Narrow the Container with CSS

```
@media (min-width: 768px) {
  .container {
    max-width: 730px;
  }
}
```

The key things to notice about this CSS are that it's inside a media query—`@media` `(min-width: 768px)`—and rather than setting a specific width, it sets the `max-width` for the element. This allows the element to continue to flex within the page in medium and large devices, but it won't get any larger than 730px wide. This keeps the narrow design while still being responsive.

Summary

This hour covered a lot of components and design features. You learned about labels and badges as ways to include additional information about your content. You also learned about panels and wells as a way to highlight content with background colors, borders, and some indenting. Finally, you learned how to use the Jumbotron to create a large but lightweight feature box on your site.

Table 6.1 covers all the CSS classes learned in this hour.

TABLE 6.1   CSS Classes for Labels, Badges, Panels, Wells, and the Jumbotron

Class	Description
.label	Defines the element as a label
.label-default	Styles the label element with the default label style
.label-primary	Styles the label element with the primary label style
.label-success	Styles the label element with the success label style
.label-info	Styles the label element with the info label style
.label-warning	Styles the label element with the warning label style
.label-danger	Styles the label element with the danger label style
.badge	Defines the element as a badge
.well	Defines the element as a well
.well-lg	Styles the well element as a large well
.well-sm	Styles the well element as a small well
.panel	Defines the element as a panel
.panel-default	Styles the panel with the default panel style
.panel-primary	Styles the panel with the primary panel style
.panel-success	Styles the panel with the success panel style
.panel-info	Styles the panel with the info panel style
.panel-warning	Styles the panel with the warning panel style
.panel-danger	Styles the panel with the danger panel style
.panel-heading	Defines the element as a panel heading
.panel-body	Defines the element as a panel body
.panel-footer	Defines the element as a panel footer
.panel-title	Defines the element as a panel title
.jumbotron	Defines the element as a Jumbotron

Workshop

The workshop contains quiz questions to help you process what you've learned in this hour. Try to answer all the questions before you read the answers.

Q&A

Q. How do I make the badges update when the number changes?

A. Badges are often used to display automated features such as the number of new messages in an inbox. That automation is usually done with things like PHP or JavaScript. A lot of scripts are included in Bootstrap, but the actual functionality for an inbox counter or other badge feature is beyond the scope of Bootstrap. If you need this functionality, you should search the Web to find a script that meets your needs.

Q. If I don't like the default background color for wells, can I change it?

A. Yes, you change it with your style sheet. The quickest way is to add a style to the `.well` property.

```
.well {
  background-color: #white;
}
```

Q. Does using a `.jumbotron` class change the styles of the contained content?

A. For the most part, no. But a few styles are added. For instance, paragraphs inside the Jumbotron are given a 15px margin, 21px font size, and 200 font weight. If an `<hr>` tag is in the Jumbotron, it will be given a border top color. And, as with all other Bootstrap styles, you can add your own style sheet to create your own designs.

Quiz

1. How do you add a label to your content?

   a. Add the `.label` class to any element.

   b. Add the `.label` and `.label-[variation]` classes to any element.

   c. Add the `#label` ID to any element.

   d. Add the `data-label="label content"` attribute to the element to be labeled.

2. What's the difference between a label and a badge?

   a. There is no difference.

   b. A label is text and a badge is a number.

   c. Labels are primarily text, while badges are more iconographic.

   d. Labels are primarily iconographic, while badges are more text focused.

3. True or False: You should never create your own style classes when modifying Bootstrap.

4. Which of the following is not a label or panel variation?

    a. main

    b. default

    c. danger

    d. success

5. Why should you leave badges empty rather than using "0" or "null"?

    a. So that the badges don't look odd on the page.

    b. So that the badges disappear when they are empty.

    c. So that the badges don't get mistaken for labels.

    d. There is no reason to not use "0" or "null" in badges.

6. True or False: Wells and blockquotes are displayed the same way in Bootstrap.

7. Which of these is a valid well class?

    a. .well-lg

    b. .well-default

    c. .well-info

    d. All of the above

8. True or False: Panel variations don't change the panel if there is no heading.

9. True or False: This panel has a title.

```
<div class="panel panel-default">
   <div class="panel-heading">
     This content is in a panel header.
   </div>
   <div class="panel-body">
     This content is in a panel.
   </div>
   <div class="panel-footer">
     This is content in a panel footer.
   </div>
</div>
```

10. True or False: Jumbotrons only work within a .container class.

Quiz Answers

1. b. Add the `.label` and `.label-[variation]` classes on the element containing the label content.

2. c. Labels are primarily text, while badges are more iconographic.

3. False. A good way to modify how Bootstrap displays is by creating your own style classes and applying those to your design.

4. a. There are six variations in Bootstrap: default, primary, info, success, warning, and danger.

5. b. Badges that are empty will disappear automatically.

6. False. Although both wells and blockquotes are typically displayed indented, Bootstrap displays them slightly differently.

7. a. The `.well-lg` class defines a large well.

8. True. Panel variations don't change the panel if there is no heading.

9. False. Although it has a heading, it does not have a title set with the `.panel-title` class.

10. False. Jumbotrons work within a `.container` class or outside of it. But they do look different.

Exercises

1. Find a portion of your website that could use a label or badge, and add it to the content. Remember that badges are often more dynamic than labels.

2. Put some content inside a well or panel. An easy way to decide which to use is to look at whether the content needs a header (or footer). If it doesn't need those elements, then a well might be all you need.

3. Create a Jumbotron layout. Find something you'd like to feature and put it in a Jumbotron to give it focus for that page.

HOUR 7
Bootstrap Typography

What You'll Learn in This Hour:

▶ How Bootstrap uses typography
▶ How to create headlines and headers
▶ How to modify body copy typography
▶ How Bootstrap adjusts text elements

Typography is an important part of web designs because web pages are primarily text. In this hour you learn the default Bootstrap typography styles and what things you might consider changing to meet the needs of your websites.

This hour covers the basic fonts and forms Bootstrap uses by default and then goes into detail about how to use headings to create better headlines and header elements, how to adjust the main copy on your web pages, and how to add alignments, colors, and transformations. Finally, you learn how Bootstrap adjusts the look of some HTML elements such as `<blockquote>`, `<code>`, `<abbr>`, `<ul>`, `<ol>`, and `<address>`.

Basic Typography in Bootstrap

Bootstrap takes care of basic typography for you. This means you don't need to worry about which font family (or font stack) to use. You don't have to adjust font sizes for readability, and you don't have to fiddle with line heights. Table 7.1 shows the default typography styles Bootstrap uses.

TABLE 7.1 Default Bootstrap Typography for All Media Types

Element	Font Stack	Font Size	Line Height
html	sans-serif	10px	
body	Helvetica Neue, Helvetica, Arial, sans-serif	14px	1.42857143
All headlines	Inherited		1.1

Element	Font Stack	Font Size	Line Height
h1		36px	
h2		30px	
h3		24px	
h4		18px	
h5		14px	
h6		12px	
small element inside any headline	inherited		1
h1 small, h2 small, h3 small		65%	
h4 small, h5 small, h6 small		75%	
.lead	inherited	16px	1.4

What this means is that text on the page will start out in a sans-serif font at 10px, but because Bootstrap pages must have a `<body>` tag, they will immediately be changed to the font stack: `Helvetica Neue, Helvetica, Arial, sans-serif`. In CSS, a font stack is a list of fonts ordered by design preference. If the computer viewing the page does not have the first font family, it moves to the second, and so on. The last font on the list should always be a default font family such as `sans-serif` or `monospace`.

CAUTION

Helvetica Neue May Render Incorrectly

Helvetica Neue is not a default system font on all computers, and on computers where it can be found it often isn't fully installed. This means that your pages might display with strange characters or messed-up font sizes. If you are noticing this problem, you should add a line to your custom style sheet to override the font stack in Bootstrap:

```
font-family: Arial,Helvetica,"Helvetica Neue",sans-serif !important;
```

These styles are defined in the style sheet first, so they will apply to all device sizes by default. There are some other adjustments that Bootstrap makes for larger screens, such as giving `<small>` elements and the .lead class different font sizes, but if you understand basic typography, the adjustments for larger devices won't be surprising.

Bootstrap 3 Font Sizes Are Not Best Practices

One complaint about Bootstrap typography is that it doesn't use relative font sizes, such as em and rem. It uses absolute pixel sizes for all the typography, and this can cause accessibility and design problems. But these font size measures are not well supported in Internet Explorer 8. There will be support for rems in version 4 of Bootstrap when they remove Internet Explorer 8 support. Until then, if you need to use ems or rems, you'll need to add them to your custom style sheets.

Headings

Most web pages use headings, and Bootstrap provides default styles for headings. Bootstrap adds

- ▶ Styles for headlines

- ▶ Styles for secondary or embedded subheads

- ▶ A header component to call out headlines as document headers

Headlines

Bootstrap provides default font sizes and line heights for all the standard HTML headings: <h1>, <h2>, <h3>, <h4>, <h5>, and <h6>. It also provides the same styles as classes so you can add inline headings to your documents. Figure 7.1 shows you how Bootstrap headlines look by default, and Listing 7.1 shows you how the HTML is written.

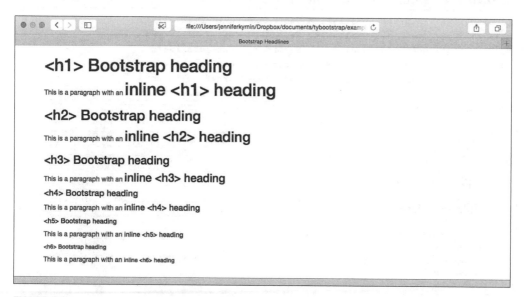

FIGURE 7.1
Bootstrap headlines.

LISTING 7.1   HTML for Bootstrap Headlines

```
<h1>&lt;h1&gt; Bootstrap heading</h1>
<p>This is a paragraph with an <span class="h1">inline &lt;h1&gt;
heading</span></p>
<h2>&lt;h2&gt; Bootstrap heading</h2>
<p>This is a paragraph with an <span class="h2">inline &lt;h2&gt;
heading</span></p>
<h3>&lt;h3&gt; Bootstrap heading</h3>
<p>This is a paragraph with an <span class="h3">inline &lt;h3&gt;
heading</span></p>
<h4>&lt;h4&gt; Bootstrap heading</h4>
<p>This is a paragraph with an <span class="h4">inline &lt;h4&gt;
heading</span></p>
<h5>&lt;h5&gt; Bootstrap heading</h5>
<p>This is a paragraph with an <span class="h5">inline &lt;h5&gt;
heading</span></p>
<h6>&lt;h6&gt; Bootstrap heading</h6>
<p>This is a paragraph with an <span class="h6">inline &lt;h6&gt;
heading</span></p>
```

One nice feature Bootstrap includes is the ability to create an inline subheader on all your
headlines using the <small> tag. This will create a lighter, inline block that complements the
headline. Listing 7.2 shows how to do this, and Figure 7.2 shows what it looks like.

LISTING 7.2   Secondary Text in Headlines

```
<h1>&lt;h1&gt; Headline <small>With Secondary Text</small></h1>
<h2>&lt;h2&gt; Headline <small>With Secondary Text</small></h2>
<h3>&lt;h3&gt; Headline <small>With Secondary Text</small></h3>
<h4>&lt;h4&gt; Headline <small>With Secondary Text</small></h4>
<h5>&lt;h5&gt; Headline <small>With Secondary Text</small></h5>
<h6>&lt;h6&gt; Headline <small>With Secondary Text</small></h6>
```

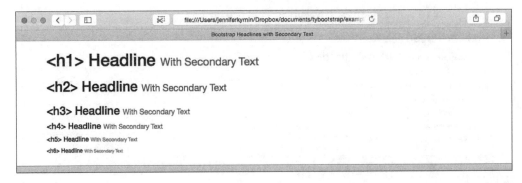

FIGURE 7.2
Bootstrap headlines with secondary text.

Bootstrap also provides the `.small` class to add secondary text to inline headlines, as in Listing 7.3 and Figure 7.3.

LISTING 7.3   Secondary Headlines on Inline Headlines

```
<p>This is a paragraph with a <span class="h1">inline &lt;h1&gt;
heading <span class="small">and secondary headline</span></span>
Lorem ipsum dolor sit amet, consectetur adipiscing elit. </p>
<p>This is a paragraph with a <span class="h2">inline &lt;h2&gt;
heading <span class="small">and secondary headline</span></span>
Lorem ipsum dolor sit amet, consectetur adipiscing elit. </p>
<p>This is a paragraph with a <span class="h3">inline &lt;h3&gt;
heading <span class="small">and secondary headline</span></span>
Lorem ipsum dolor sit amet, consectetur adipiscing elit. </p>
<p>This is a paragraph with a <span class="h4">inline &lt;h4&gt;
heading <span class="small">and secondary headline</span></span>
Lorem ipsum dolor sit amet, consectetur adipiscing elit. </p>
<p>This is a paragraph with a <span class="h5">inline &lt;h5&gt;
heading <span class="small">and secondary headline</span></span>
Lorem ipsum dolor sit amet, consectetur adipiscing elit. </p>
<p>This is a paragraph with a <span class="h6">inline &lt;h6&gt;
heading <span class="small">and secondary headline</span></span>
Lorem ipsum dolor sit amet, consectetur adipiscing elit. </p>
```

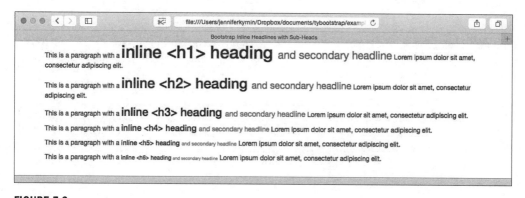

FIGURE 7.3
Secondary headlines on inline headlines.

But you aren't limited to text headlines in Bootstrap. With the `.text-hide` class, you can create a graphic headline and replace the text while keeping the page accessible.

▼ TRY IT YOURSELF

Replace a Headline with an Image

Bootstrap lets you easily replace text headlines with images. In just a couple of steps you can create a custom headline that is also accessible and SEO-friendly:

1. Open your Bootstrap page in an HTML editor.

2. Add an `<h1>` headline:

   ```
   <h1>Headline</h1>
   ```

3. Add the class `.text-hide` to the headline:

   ```
   <h1 class="text-hide">
   ```

4. Add the ID `#mainhead` to the headline:

   ```
   <h1 class="text-hide" id="mainhead">
   ```

5. Add the following CSS to your style sheet:

   ```
   #mainhead {
       background-image: url(images/headline.png);
       background-repeat: no-repeat;
       width:458px;
       height:76px;
   }
   ```

Be sure to point to your own image and set the width and height to the correct size, or use background image styles to clip and resize the image to fit your design. Listing 7.4 shows the HTML used to create the page in Figure 7.4.

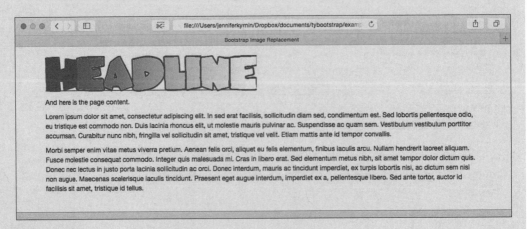

FIGURE 7.4
The headline is an `<h1>` replaced by an image.

LISTING 7.4   Image Replacement with Bootstrap

```
<!DOCTYPE html>
<html lang="en">
<head>
    <meta charset="utf-8">
    <meta http-equiv="X-UA-Compatible" content="IE=edge">
    <meta name="viewport"
      content="width=device-width, initial-scale=1">
    <title>Bootstrap Image Replacement</title>

    <!-- Bootstrap -->
    <link href="css/bootstrap.min.css" rel="stylesheet">
    <!-- HTML5 shim and Respond.js for IE8 support of HTML5
    elements and media queries -->
    <!-- WARNING: Respond.js doesn't work if you view the page
    via file:// -->
    <!--[if lt IE 9]>
      <script
src="https://oss.maxcdn.com/html5shiv/3.7.2/html5shiv.min.js">
</script>
      <script
src="https://oss.maxcdn.com/respond/1.4.2/respond.min.js"></script>
    <![endif]-->
    <style>
      #mainhead {
        background-image: url(images/headline.png);
        background-repeat: no-repeat;
        width:458px;
        height:76px;
      }
    </style>
  </head>
  <body>
  <div class="container">
    <h1 class="text-hide" id="mainhead">Headline</h1>
    <p>And here is the page content.</p>
    <p>Lorem ipsum dolor sit amet, consectetur adipiscing elit.
    In sed erat facilisis, sollicitudin diam sed, condimentum est.
    Sed lobortis pellentesque odio, eu tristique est commodo non.
    Duis lacinia rhoncus elit, ut molestie mauris pulvinar ac.
    Suspendisse ac quam sem. Vestibulum vestibulum porttitor
    accumsan. Curabitur nunc nibh, fringilla vel sollicitudin sit
    amet, tristique vel velit. Etiam mattis ante id tempor
    convallis.</p>
    <p>Morbi semper enim vitae metus viverra pretium. Aenean felis
    orci, aliquet eu felis elementum, finibus iaculis arcu. Nullam
```

```
        hendrerit laoreet aliquam. Fusce molestie consequat commodo.
        Integer quis malesuada mi. Cras in libero erat. Sed elementum
        metus nibh, sit amet tempor dolor dictum quis. Donec nec lectus
        in justo porta lacinia sollicitudin ac orci. Donec interdum,
        mauris ac tincidunt imperdiet, ex turpis lobortis nisi, ac
        dictum sem nisi non augue. Maecenas scelerisque iaculis
        tincidunt. Praesent eget augue interdum, imperdiet ex a,
        pellentesque libero. Sed ante tortor, auctor id facilisis sit
        amet, tristique id tellus.</p>
    </div>
    </body>
</html>
```

Headers

Bootstrap adds a component to help you call out the header on a web page. To use it, you add the `.page-header` class to a container element surrounding the `<h1>` element that titles the entire page.

This creates a box around the headline with a 9px bottom padding, adds more space around the top and bottom of the box, and adds a bottom border. Figure 7.5 shows what a header box looks like.

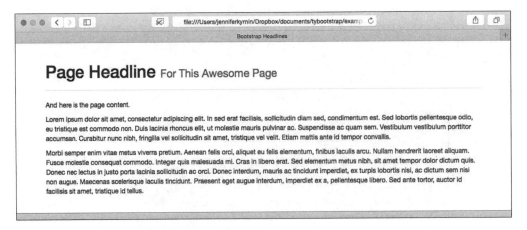

FIGURE 7.5
A simple page with a header box around the headline.

The HTML for this is easy to implement, as Listing 7.5 shows.

LISTING 7.5   Adding a Page Header

```
<div class="page-header">
  <h1>Page Headline <small>For This Awesome Page</small></h1>
</div>
```

Body Copy Text

Body copy is the majority of text on a web page. It is text that is found inside paragraphs and divisions of text. With Bootstrap, you can put body copy straight into the `<body>` of the document, include it in `<div>` containers, or put it inside `<p>` tags.

The default font is, as mentioned in Table 7.1, 14px tall; uses the font stack `Helvetica Neue, Helvetica, Arial, sans-serif`; and has a line height of about 1.4.

NOTE

Line Height Doesn't Require a Measurement Unit

You will often see people define the line height as 14px or 2em, but this isn't required. If you use a value like 1.4, this tells the browser to create a line height that is 1.4 times the computed font size. Then if the browser font is resized in some way, the line height will remain 1.4 times the font size.

Bootstrap gives paragraphs a small typography boost by adding a bottom margin of half the computed line height. This adds a bit of space between paragraphs and other elements, making the text easier to read. As shown in Figure 7.6, the first paragraph does not have any space between it and the second because it is not inside `<p>` tags.

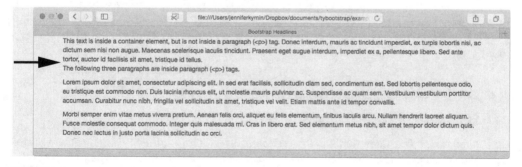

FIGURE 7.6
Blocks of text in paragraphs and not in paragraphs.

Best practices recommend that you always include blocks of text inside paragraph tags. This will keep all your text as legible as possible.

Bootstrap also lets you highlight lead copy with the `.lead` class. This creates a block of text that is slightly larger than the standard body copy. You can use it on a paragraph or a few words to make them stand out.

▼ TRY IT YOURSELF

Add a Lead to Some Copy

A lead is often highlighted in some way to make it stand out from the main text. Bootstrap lets you do this with the `.lead` class. Plus, you can apply it to both entire paragraphs or just a few words of the text:

1. Open your Bootstrap web page in an HTML editor.

2. Go to the first paragraph that you want to highlight, and add `class="lead"` to the paragraph tag:

```
<p class="lead">
```

3. View the page in a web browser and see that the first paragraph is larger than subsequent paragraphs. You can see this in Figure 7.7.

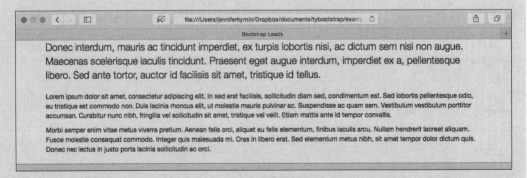

FIGURE 7.7
A page with a lead paragraph.

You can also make just the first three words the lead.

4. Open the page in your HTML editor.

5. Surround the first three words in the first paragraph with a `<span></span>` element. Note: if you're editing the same page, remove the `class="lead"` from the `<p>` tag.

6. Add `class="lead"` to the `<span>` tag and view in your browser. Figure 7.8 shows how this looks.

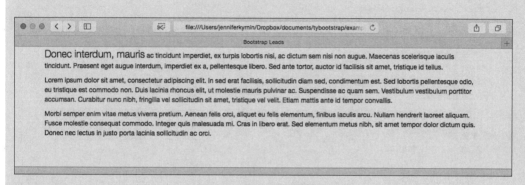

FIGURE 7.8
A page with lead text in the first paragraph.

Inline Text

A number of HTML elements are available that you can use to define different aspects of the text. Bootstrap styles these elements for you. The elements are

- ▶ `<del>`—The text has been deleted from the document.

- ▶ `<em>`—The text is emphasized in italics.

- ▶ `<ins>`—The text has been inserted into the document.

- ▶ `<mark>`—The text is highlighted for reference purposes.

- ▶ `<s>`—The text has been struck from the document and has a line through it.

- ▶ `<small>`—The text is in small print such as legalese.

- ▶ `<strong>`—The text is emphasized in bold.

- ▶ `<u>`—The text is underlined.

Figure 7.9 shows how Bootstrap styles these elements. If you need them to look differently, you can always add styles to your personal style sheet.

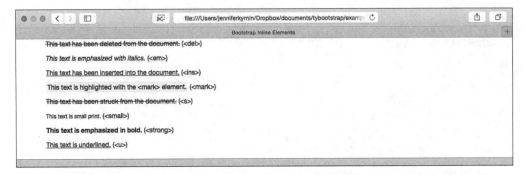

FIGURE 7.9
Inline text styles in Bootstrap.

Aligning Elements

Bootstrap offers several classes to help you align both text and block elements:

▶ `text-left`

▶ `text-center`

▶ `text-right`

▶ `text-justify`

▶ `text-nowrap`

▶ `pull-left`

▶ `pull-right`

▶ `center-block`

CAUTION

Don't Use the `.pull-*` Classes on Navigation Elements

Specific classes for navigation elements let you float navigation to the right or left that are optimized for navigation. This is covered in more detail in Hour 12, "Creating Navigation Systems with Bootstrap."

The `.text-left`, `.text-right`, and `.text-center` classes do exactly what they appear to do and position text on the left, right, or center sides of the block in which they are contained. You can use the `.pull-right`, `.pull-left`, and `.center-block` classes in the same way on block elements as entire paragraphs. The `.text-justify` class justifies the text across both sides of the block, and the `.text-nowrap` class forces the text to ignore typical wrapping and scroll horizontally. Figure 7.10 demonstrates these styles.

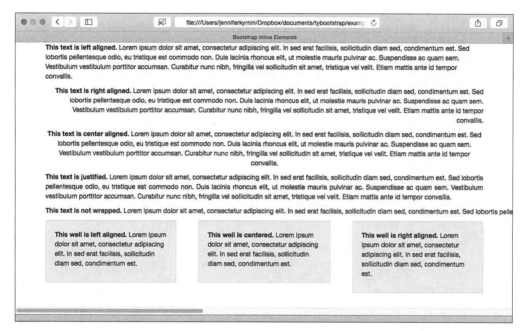

FIGURE 7.10
Alignment styles in Bootstrap.

Transforming Text Elements

Bootstrap offers several classes to let you transform text elements. Figure 7.11 shows text that has been transformed with these classes:

- ▶ `text-capitalize`—Capitalizes the first letter of every word
- ▶ `text-lowercase`—Makes every letter lowercase
- ▶ `text-uppercase`—Makes every letter uppercase

FIGURE 7.11
Transformation styles in Bootstrap.

In addition, standard Bootstrap helper classes are available that you can add to any element to change the text and background colors:

- ▶ text-muted
- ▶ text-primary
- ▶ text-success
- ▶ text-info
- ▶ text-warning
- ▶ text-danger
- ▶ bg-primary
- ▶ bg-success
- ▶ bg-info
- ▶ bg-warning
- ▶ bg-danger

These classes change either the text color (.text-*) or the background color (.bg-*) on the elements. You can use them to visually convey specific information, but be sure to provide other contextual clues beyond the color if the classes impart important or relevant information. Figure 7.12 shows how these styles look.

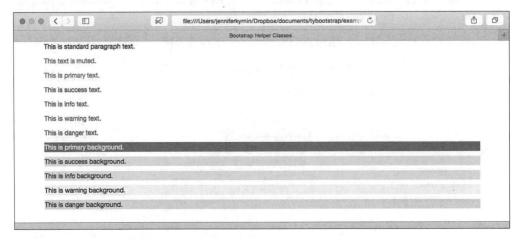

FIGURE 7.12
Helper classes change the text and background colors.

Other Text Blocks

HTML offers many other types of text, including code samples, quotations, lists, abbreviations, and addresses. Bootstrap has special style options for each.

Code

The following are five HTML tags you can use to define code, and Bootstrap has styles for all of them:

▶ `<code>`—Defines inline code samples inside paragraphs of text

▶ `<kbd>`—Defines user input, typically entered via a keyboard

▶ `<pre>`—Defines multiple lines of code, separated into a block

▶ `<samp>`—Defines sample output from a computer code

▶ `<var>`—Defines variables

Figure 7.13 shows how these look on a Bootstrap page.

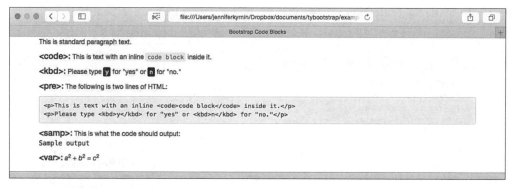

FIGURE 7.13
Bootstrap code blocks.

You can add the class `.pre-scrollable` to the `<pre>` tag to set a maximum height on the `<pre>` blocks of 350px and add a scrollbar to see any additional text.

Quotations

Bootstrap uses the HTML tag `<blockquote>` to define long quotations. Best practices in Bootstrap are to use the `<blockquote>` tag around any quotation that itself has HTML tags around it. If the quote is short enough to be written inline, then you should just put quotation marks around it.

If you need to name a source for your quotation, include the quotation in a paragraph and then wrap the source in a `<footer>` tag inside the `<blockquote>`. Include the `<cite>` tag around the actual source. Listing 7.6 shows how.

LISTING 7.6   A Quotation with Source

```
<blockquote>
<p>`Twas brillig, and the slithy toves<br>
Did gyre and gimble in the wabe:<br>
All mimsy were the borogoves,<br>
And the mome raths outgrabe.</p>
<footer>by Lewis Carroll from <cite title="The Jabberwocky">"The
Jabberwocky"</cite></footer>
</blockquote>
```

Bootstrap makes the font larger and adds a gray border to the left side of standard quotations. The footer is then slightly smaller and in light gray text. If you want your quotation to be right-aligned, you can use the class `.blockquote-reverse` to align the quote to the right and move the border to the right side. Figure 7.14 shows what blockquotes look like in Bootstrap.

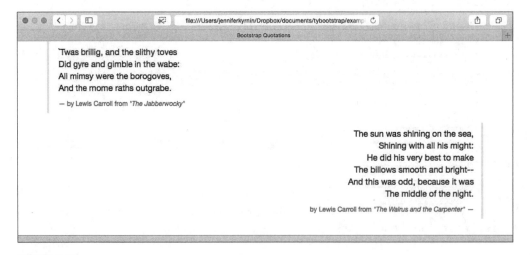

FIGURE 7.14
Bootstrap quotations.

Lists

HTML has three types of lists: unordered lists, ordered lists, and definition lists. Bootstrap styles these with similar typography as the body copy. But there are also some extra classes to adjust the styles of the lists:

- ▶ `.list-unstyled`—Removes the default styles and left margin on list items.

- ▶ `.list-inline`—Places all the list items on a single line with a little padding between.

- ▶ `.dl-horizontal`—This places the terms and descriptions in a definition list lined up side-by-side.

Figure 7.15 shows these list styles in a typical Bootstrap page.

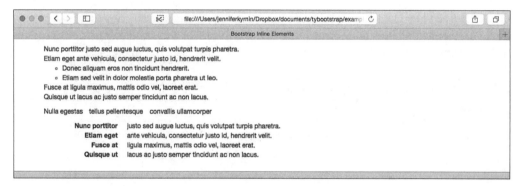

FIGURE 7.15
Bootstrap list styles.

Abbreviations

Bootstrap adds some default styles to the standard HTML `<abbr>` tag: a change to the cursor and a dotted line underneath the abbreviation. If you include the definition in the `title` attribute, most browsers will display that when the customer hovers over the element.

However, when you use abbreviations that are all caps, such as "NASA" and "HTML," the capital letters can make the text look larger than it is. Bootstrap provides a class, `.initialism`, to display the abbreviation in a slightly smaller font size. Listing 7.7 shows how you might use this class.

LISTING 7.7   Using the `.initialism` Class on an Abbreviation

```
<p>I wrote a book on <abbr title="HyperText Markup Language 5"
class="initialism">HTML5</abbr> and another on <abbr
title="Responsive Web Design" class="initialism">RWD</abbr>.</p>
```

Remove the class from one of the two abbreviations to see the difference.

Addresses

HTML also provides a tag, `<address>`, to present contact information either about the preceding element or the document as a whole. Bootstrap adds a larger bottom margin and resets the font style and line height.

Summary

This hour covered a lot of Bootstrap styles that help adjust the typography of your web pages. You learned about the basic fonts, font sizes, and line heights Bootstrap assigns. You also learned how to create headlines and page headers.

Bootstrap styles many HTML elements that appear in the body copy. This hour covered more than 20 classes you can use to adjust the way different portions of the body copy look.

You also learned how Bootstrap styles text elements such as `<code>`, `<blockquote>`, `<abbr>`, `<address>`, and HTML lists (`<ul>`, `<ol>`, and `<dl>`). Plus, this hour covered some classes specific for those elements. Table 7.2 explains all the classes Bootstrap adds for typography.

TABLE 7.2 Bootstrap Typography Classes

CSS Class	Description
`.bg-danger`	Changes the background color to indicate the element is "danger"
`.bg-info`	Changes the background color to indicate the element is "info"
`.bg-primary`	Changes the background color to indicate the element is "primary"
`.bg-success`	Changes the background color to indicate the element is "success"
`.bg-warning`	Changes the background color to indicate the element is "warning"
`.blockquote-reverse`	Moves the quotation to be right aligned
`.center-block`	Centers the block element
`.dl-horizontal`	Converts the definition list to be horizontal
`.h1, .h2, .h3, .h4, .h5, .h6`	Creates inline headlines
`.initialism`	Makes an abbreviation a slightly smaller font
`.lead`	Makes the text larger to be a lead for the page
`.list-inline`	Creates an inline list
`.list-unstyled`	Removes the left margin and list styles
`.page-header`	Creates a header section on the page
`.pre-scrollable`	Sets the `max-height` of a `<pre>` block to 350px and adds a scrollbar.

CSS Class	Description
.pull-left	Floats a display element left
.pull-right	Floats a display element right
.small	Makes the font slightly smaller
.text-capitalize	Transforms the text to capitalize the words
.text-center	Centers the text
.text-danger	Colors the text to indicate danger
.text-hide	Hides the text to create image replacements
.text-info	Colors the text to indicate information
.text-justify	Justifies the text
.text-left	Aligns the text to the left
.text-lowercase	Transforms the text to all lowercase
.text-muted	Colors the text to indicate it's muted
.text-nowrap	Turns off wrapping on the text
.text-primary	Colors the text to indicate it's primary
.text-right	Aligns the text to the right
.text-success	Colors the text to indicate it's success
.text-uppercase	Transforms the text to all uppercase
.text-warning	Colors the text to indicate it's a warning

Workshop

The workshop contains quiz questions to help you process what you've learned in this hour. Try to answer all the questions before you read the answers.

Q&A

Q. When I apply the `.pull-right` or `.pull-left` classes to my elements, they still take up the full width of the page. How do I fix that?

A. HTML automatically fills up all the horizontal space available unless you set a width. To fix this, add a width style to your style sheet for that element. For example,

```
<div id="pullRight" class="pull-right"> ... </div>

<style>
  #pullRight {
    width: 30%;
  }
</style>
```

Q. Do I have to use the colors that are defined for the helper classes, like `.bg-warning` and `.text-info`?

A. You can change them with your personal style sheets or with Less mixins. You'll learn more about that in Hour 23, "Using Less and Sass with Bootstrap."

Quiz

1. What happens if a browser doesn't have Helvetica Neue?

    a. The page displays in Helvetica.

    b. The page displays in the font the customer likes.

    c. The page displays in a random font.

    d. The page doesn't display.

2. What does it mean that a font property is inherited?

    a. The font property uses what the browser suggests.

    b. The font property is defined by the parent element.

    c. The font property stays the same.

    d. The font property is defined by a personal style sheet.

3. What is the pixel size of the line-height of body copy in all devices?

    a. The same as the body copy font size.

    b. 1.42857143 pixels.

    c. 1.4px.

    d. Line height is not defined in pixels.

4. How do you define headlines in Bootstrap?

    a. With the `<h1>` through `<h6>` tags

    b. With the `.h1` through `.h6` classes

    c. With the `<headline>` tag

    d. Both a and b

    e. All of the above

5. What does the `.lead` class do?

    a. Adds extra leading to the text

    b. Highlights the text with bold and colors

    c. Makes the text larger

    d. Makes the text smaller

6. How do you center paragraph text?

    a. Use the `<center>` tag.

    b. Use the `.center` class.

    c. Use the `.text-center` class.

    d. Use the `.block-center` class.

7. How do you make text uppercase in Bootstrap?

    a. Use the `.text-uppercase` class.

    b. Use the CSS property `text-transform`.

    c. Use the `<uppercase>` tag.

    d. You can't make text uppercase with Bootstrap.

8. Which of these is not a Bootstrap helper class?

    a. `.text-danger`

    b. `.text-info`

    c. `.bg-danger`

    d. `.background-danger`

9. Which of these is not a code HTML tag that Bootstrap styles?

    a. `<code>`

    b. `<kbd>`

    c. `<tt>`

    d. `<var>`

10. Why use the `.initialism` class?

    a. To make your text slightly smaller.

    b. To make all caps abbreviations appear more the same size as the surrounding text.

    c. To create an initial cap.

    d. It is not a valid Bootstrap class.

Quiz Answers

1. a. The next family in the font stack is Helvetica, so that is what the browser will use.

2. b. The font property is set by the parent element.

3. d. The line height is defined as a unitless number (1.42857143).

4. d. You can define headlines with both the `<h1>` through `<h6>` tags and the `.h1` through `.h6` classes.

5. c. The `.lead` class makes the text larger.

6. c. The `.text-center` class will center the paragraph text inside a block.

7. a. Use the Bootstrap `.text-uppercase` class.

8. d. The `.background-danger` class is not a Bootstrap helper class.

9. c. The `<tt>` tag is not styled by Bootstrap.

10. b. The `.initialism` class resizes the font to make the text more uniform.

Exercises

1. Open your Bootstrap web page and add a list to the content. Try the different list classes to see whether there are any that work well in your design.

2. Create a blockquote in your page to highlight a quotation.

Styling Tables

What You'll Learn in This Hour:

▶ How Bootstrap styles tables

▶ The Bootstrap table classes

▶ How Bootstrap panels interact with tables

▶ How Bootstrap makes tables responsive

Tables are an important part of web pages because they provide a way to display tabular data effectively. But tables can be difficult to handle responsively. Bootstrap provides lots of styles and classes to create data tables that look nice and are responsive.

Basic Tables

Bootstrap applies three styles to HTML tables automatically:

▶ `background-color: transparent;`—The background color is transparent.

▶ `border-spacing: 0;`—The border spacing is set to zero (0).

▶ `border-collapse: collapse;`—The borders are collapsed.

Bootstrap styles other table tags, including `<caption>`, `<th>`, and `<td>`.

To get the most out of Bootstrap tables, you should get in the habit of using the optional `<thead>`, `<tbody>`, and `<tfoot>` tags when appropriate, and always use the `.table` class on your tables. This will define the width of your table to 100% of the screen and is required to use some of the advanced classes discussed later in this hour.

Code Listing 8.1 shows a standard Bootstrap table in HTML.

LISTING 8.1   A Basic Bootstrap Table

```
<table class="table">
  <caption>Contact Information</caption>
  <thead>
    <tr>
      <th>Name</th>
      <th>Title</th>
      <th>URL</th>
      <th>Email</th>
    </tr>
  </thead>
  <tbody>
    <tr>
      <td>Jennifer Kyrnin</td>
      <td>Chief Dandylion Officer</td>
      <td>http://htmljenn.com/</td>
      <td>htmljenn@gmail.com</td>
    </tr>
    <tr>
      <td>McKinley</td>
      <td>Dandelion Observation Officer</td>
      <td>http://responsivewebdesignin24hours.com/mckinley</td>
      <td>mckinley@rwdin24hours.com</td>
    </tr>
    <tr>
      <td>Rambler</td>
      <td>Chief Taste Tester</td>
      <td>http://responsivewebdesignin24hours.com/rambler</td>
      <td>rambler@rwdin24hours.com</td>
    </tr>
  </tbody>
  <tfoot>
    <tr>
      <td colspan="4"><p>McKinley and Rambler are a dog and a
      horse, respectively.</p></td>
    </tr>
  </tfoot>
</table>
```

If you forget the `.table` class, your table can appear very crowded and difficult to read. Figure 8.1 shows the same table, both with and without the `.table` class.

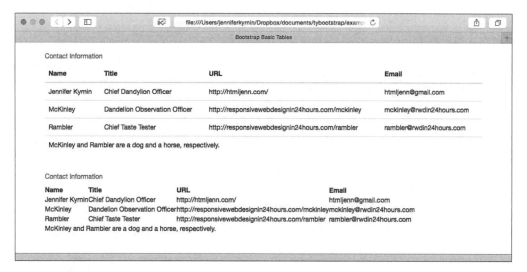

FIGURE 8.1
A Bootstrap table with and without the `.table` class.

Bootstrap Table Classes

Bootstrap provides more styling of tables than just the default table styles. There are several classes you can add to your tables to add more features that more sophisticated tables use:

- ▶ `.table-striped`—Adds zebra-striped rows within the `<tbody>` tag of your table.

- ▶ `.table-bordered`—Adds borders around all sides (not just the bottom) of the table and between the rows and columns.

- ▶ `.table-hover`—Enables a hover state on rows within the `<tbody>`.

- ▶ `.table-condensed`—Cuts the padding in half to make the table more compact.

You add these classes to the `<table>` tag, as in Listing 8.2.

LISTING 8.2   Table Classes in Bootstrap

```
<table class="table table-bordered table-striped table-hover
table-condensed">
```

You can use any combination of the classes to adjust your table how you like. Figure 8.2 shows a table using all the fields.

FIGURE 8.2
A Bootstrap table with borders, stripes, and condensed.

CAUTION

Striped Rows Don't Work in Internet Explorer 8

The row stripes are added to the rows with the :nth-child selector in CSS. However, Internet Explorer does not support the :nth-child selector. If you really need this effect in Internet Explorer 8, you can check out the JavaScript utility Selectivizr (http://selectivizr.com/).

▼ TRY IT YOURSELF

Adjust the Zebra Stripes on Your Table

The zebra stripes are really useful for making tables much more legible, but the default colors are very bland. It is possible to change the colors to match your color scheme:

1. Open your web page in your HTML editor.

2. Make sure your table has both the .table and .table-striped classes on it.

3. Add a style sheet to the <head> of the document.

4. Add the style rule .table-striped>tbody>tr:nth-of-type(odd).

5. Change the background color with the style property background-color: #d6adfa;.

6. Make sure the font color works with the background color. If the contrast doesn't work, change it with the color property.

Listing 8.3 shows the full HTML, and Figure 8.3 shows how it looks in a browser.

LISTING 8.3 Change the Zebra Stripe Colors

```
<!DOCTYPE html>
<html lang="en">
  <head>
    <meta charset="utf-8">
    <meta http-equiv="X-UA-Compatible" content="IE=edge">
    <meta name="viewport"
    content="width=device-width, initial-scale=1">
    <title>Bootstrap Table Classes</title>

    <!-- Bootstrap -->
    <link href="css/bootstrap.min.css" rel="stylesheet">
    <!-- HTML5 shim and Respond.js for IE8 support of HTML5
    elements and media queries -->
    <!-- WARNING: Respond.js doesn't work if you view the page
    via file:// -->
    <!--[if lt IE 9]>
      <script
src="https://oss.maxcdn.com/html5shiv/3.7.2/html5shiv.min.js">
      </script>
      <script
src="https://oss.maxcdn.com/respond/1.4.2/respond.min.js"></script>
    <![endif]-->
    <style>
      .table-striped>tbody>tr:nth-of-type(odd) {
        background-color: #d6adfa;
      }
    </style>
  </head>
  <body>
  <div class="container">
    <p>
    <table class="table table-striped">
      <caption>Contact Information</caption>
      <thead>
        <tr>
          <th>Name</th>
          <th>Title</th>
          <th>URL</th>
          <th>Email</th>
        </tr>
      </thead>
      <tbody>
        <tr>
          <td>Jennifer Kyrnin</td>
```

```
              <td>Chief Dandylion Officer</td>
              <td>http://htmljenn.com/</td>
              <td>htmljenn@gmail.com</td>
            </tr>
            <tr>
              <td>McKinley</td>
              <td>Dandelion Observation Officer</td>
              <td>http://responsivewebdesignin24hours.com/mckinley</td>
              <td>mckinley@rwdin24hours.com</td>
            </tr>
            <tr>
              <td>Rambler</td>
              <td>Chief Taste Tester</td>
              <td>http://responsivewebdesignin24hours.com/rambler</td>
              <td>rambler@rwdin24hours.com</td>
            </tr>
          </tbody>
          <tfoot>
            <tr>
              <td colspan="4"><p>McKinley and Rambler are a dog and a
              horse, respectively.</p></td>
            </tr>
          </tfoot>
        </table>
        </p>
      </div>
      </body>
</html>
```

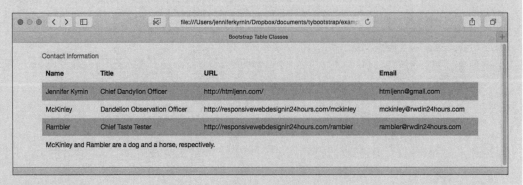

FIGURE 8.3
Adjusted zebra stripe colors.

You can also use contextual classes on your tables to add meaning to the cells or rows. You can add the following classes:

- ▶ .active—Applies the hover color

- ▶ .danger—Applies a red color to indicate a dangerous or negative action

- ▶ .info—Applies a blue color to indicate information or neutral action

- ▶ .success—Applies a green color to indicate a successful or positive action

- ▶ .warning—Applies a yellow color to indicate a warning or possible negative action

Add the class to a row of the table like this:

```
<tr class="warning">
```

Or to a specific cell on either the `<th>` or `<td>` tags:

```
<td class="success">
```

Remember that these classes change only the background color of the table elements. They do not provide any meaning. To make these accessible, you should make sure that the content conveys the meaning along with the color. For example:

```
<td class="warning">Warning: lorem ipsum sit dolor...</td>
```

You also can use the .sr-only class to define text that displays only on screen readers. For example:

```
<td class="warning"><span class="sr-only">Warning: </span>
lorem ipsum sit dolor...</td>
```

This class hides information used by screen readers.

Panels with Tables

In Hour 6, "Labels, Badges, Panels, Wells, and the Jumbotron," you learned how to add panels to your Bootstrap page. Panels combine with tables to make a more seamless feature on the page.

Any non-bordered table that is included in a panel will integrate into the panel, as in Figure 8.4. Listing 8.4 shows how simple the HTML is for this. If there is a .panel-body, an extra border will be added above the table.

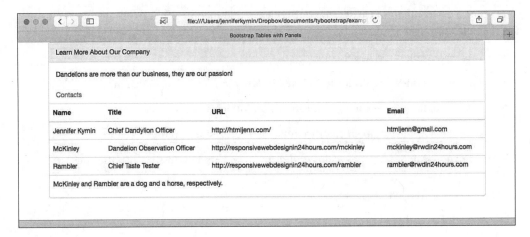

FIGURE 8.4
A table in a panel.

LISTING 8.4   A Table in a Panel

```
<div class="panel panel-default">
  <div class="panel-heading">
    Learn More About Our Company
  </div>
  <div class="panel-body">
    Dandelions are more than our business, they are our passion!
  </div>
  <table class="table">
  <caption>Contacts</caption>
  <thead>
    <tr>
      <th>Name</th>
      <th>Title</th>
      <th>URL</th>
      <th>Email</th>
    </tr>
  </thead>
  <tbody>
    <tr>
      <td>Jennifer Kyrnin</td>
      <td>Chief Dandylion Officer</td>
      <td>http://htmljenn.com/</td>
      <td>htmljenn@gmail.com</td>
    </tr>
    <tr>
      <td>McKinley</td>
      <td>Dandelion Observation Officer</td>
```

```
    <td>http://responsivewebdesignin24hours.com/mckinley</td>
    <td>mckinley@rwdin24hours.com</td>
  </tr>
  <tr>
    <td>Rambler</td>
    <td>Chief Taste Tester</td>
    <td>http://responsivewebdesignin24hours.com/rambler</td>
    <td>rambler@rwdin24hours.com</td>
  </tr>
</tbody>
<tfoot>
  <tr>
    <td colspan="4"><p>McKinley and Rambler are a dog and a
    horse, respectively.</p></td>
  </tr>
</tfoot>
</table>
</div>
```

If there is no .panel-body, then the heading will flow straight into the table.

Responsive Tables

Tables are very difficult to make responsive. They often are too wide for small screens, and many small screens don't scroll well horizontally.

Bootstrap provides a fix for that with the .table-responsive class. To make a table responsive, you should surround it with another element with the .table-responsive class, as in Listing 8.5.

LISTING 8.5   Make a Bootstrap Table Responsive

```
<div class="table-responsive">
<table class="table">
...
</table>
</div>
```

Figure 8.5 shows how a table scrolls on an iPhone.

FIGURE 8.5
A responsive table scrolling on an iPhone.

CAUTION

Bootstrap Responsive Tables Use `overflow-y: hidden`

The CSS property `overflow-y: hidden` tells the browser to clip any content that goes beyond the top or bottom edges of the table space. This can clip off dropdown menus or other widgets. So be sure to test any dynamic content you have inside responsive tables.

Bootstrap provides a basic way to make tables responsive. But several other methods are available that you can use that don't require an additional HTML tag. I cover these in more detail in my book *Sams Teach Yourself Responsive Web Design in 24 Hours*.

Summary

This hour covered how Bootstrap styles tables. It covered the basic features of Bootstrap tables. It also covered the special classes that let you dress up your tables. Table 8.1 describes all the CSS classes covered in this hour.

This hour also covered how panels adjust to contain tables, as well as a basic way to make your tables responsive.

TABLE 8.1   Bootstrap Table Classes

CSS Class	Description
.active	Changes the background of the table element to the hover color and indicates that that field is active.
.danger	Changes the background of the table element to a red color and indicates that it is a dangerous or negative action.
.info	Changes the background color of the table element to a blue color and indicates that it is an informational or neutral action.
.success	Changes the background color of the table element to a green color and indicates that it is a successful or positive action.
.table	Indicates that the table should have Bootstrap styles applied.
.table-bordered	Indicates that the table should have borders.
.table-hover	Adds hover color to the table rows or cells when readers mouse over them.
.table-responsive	When applied to a container element around a table, it will make the table scroll horizontally on smaller devices.
.table-striped	Adds zebra stripes to the rows of a table.
.warning	Changes the background color of the table element to a yellow color and indicates that it might be a danger or might have a negative action.

Workshop

The workshop contains quiz questions to help you process what you've learned in this hour. Try to answer all the questions before you read the answers.

Q&A

Q. You mention that there are other ways to make tables responsive, but I can't imagine any. What are they?

A. There are several ways you can make tables more responsive. You can hide the less critical rows or columns. You can rearrange the content so that the rows display separately, and you can resize the table cells.

Q. I never use the `<tbody>`, `<tfoot>`, and `<thead>` tags. Are they absolutely required for Bootstrap tables?

A. The `<tbody>` tag is required if you want to use any of the classes because they are applied only to the rows in the body. The `<thead>` and `<tfoot>` tags are not required.

Quiz

1. What happens if you don't use the `.table` class on your table?

    a. The table still displays with Bootstrap styles.

    b. The table displays without any styles.

    c. The table displays with borders and colors.

    d. The table displays condensed and borderless.

2. What does `.table-striped` do?

    a. Adds borders around all the cells

    b. Adds colors to all the cells

    c. Adds colors to every other row to create a zebra-striped table

    d. Does nothing

3. What does the `.table-bordered` class do?

    a. Adds borders around all the cells

    b. Adds borders between columns

    c. Adds borders between rows

    d. Does nothing

4. What does the `.table-hover` class do?

    a. Outlines cells when they are moused over

    b. Adds a background color to rows when they are moused over

    c. Changes the text color on cells when they are moused over

    d. Does nothing

5. What does the `.table-condensed` class do?

    a. Removes the cell padding on the table

    b. Cuts the cell padding in half on the table

    c. Makes the font size smaller in the table

    d. Reduces the width of the table by 50%

6. True or False: You cannot use `.table-condensed` with `.table-striped`.

7. What are table contextual classes in Bootstrap?

    a. Classes that add extra information to the table cells.

    b. Classes that change the color of the text.

    c. Classes that change the background color of table cells.

    d. There is no such thing in Bootstrap.

8. Which of these is a contextual class for tables?

    a. `.hover`

    b. `.bg-info`

    c. `.text-success`

    d. `.warning`

9. How does Bootstrap display tables in panels?

    a. The table integrates seamlessly into the panel, and a border is added to the top if there is a `.panel-body`.

    b. The table is given a new background color to make it fit in the panel.

    c. The table is minimized to fit in the panel.

    d. Bootstrap doesn't do anything special to tables in panels.

10. How do you make a Bootstrap table responsive?

    a. Add the `.table-responsive` class to the table.

    b. Add the `.responsive class` to a container `<div>` around the table.

    c. Add the `.table-responsive` class to a container `<div>` around the table.

    d. You do nothing; they are responsive by default.

Quiz Answers

1. d. The table displays condensed and borderless.

2. c. Adds colors to every other row to create a zebra-striped table.

3. a. Adds borders around all the cells.

4. b. Adds background colors to rows when they are moused over.

5. b. Cuts the cell padding in half on the table.

6. False. You can use all the table classes together on one table.

7. c. Contextual classes change the background color on table rows.

8. d. `.warning` is a table contextual class.

9. a. The table is integrated seamlessly into the panel, with an extra top border for panels with a `.panel-body`.

10. c. Add the `.table-responsive` class to a container `<div>` around the table.

Exercises

1. Add a data table to your Bootstrap web page. Add some extra classes to the table to make it look more impressive.

2. Try adding a panel heading with no panel body with a table inside. This is a great way to create a nice-looking heading for your tables.

Styling Forms

What You'll Learn in This Hour:

▶ How to create basic HTML forms

▶ How to style horizontal and inline forms with Bootstrap

▶ How to style input controls and dropdown menus

▶ How to build and use an input group

▶ How to adjust the interactivity with Bootstrap classes

HTML forms are difficult to make look nice, but Bootstrap makes it easy. In this hour you learn how to use Bootstrap classes to style basic forms. You also learn about some of the options Bootstrap provides for creating forms in different designs.

Bootstrap supports HTML5 form controls. There also are several states you can set with Bootstrap to provide more information about the form fields. This hour teaches you how to change the sizes of the controls and how to add help text to the form fields.

Basic Forms

HTML forms are easy to add to your web pages. Listing 9.1 shows you the barebones HTML you might use for a form.

CAUTION

Bootstrap Does Not Give Your Forms an Action

Bootstrap is a framework for how the web page will look. Although you can use some Bootstrap components to add interactivity to your web pages, you will need to find or develop the scripts needed to submit web forms separately. In the examples in this hour, the `<form>` tags will not have any script in the `action` attribute. If you need more help learning how to make HTML forms work, you might want to look for a book on PHP that goes into greater depth on forms.

LISTING 9.1   Basic HTML Form

```
<form action="#">
  <label for="firstName">First Name:</label>
  <input type="text" autofocus required id="firstName"
    placeholder="First Name"><br>
  <label for="lastName">Last Name:</label>
  <input type="text" required id="lastName"
    placeholder="Last Name"><br>
  <label for="email">Email:</label>
  <input type="email" required id="email"
    placeholder="Email Address"><br>
  <label for="homePhone">Home Phone:</label>
  <input type="tel" id="homePhone" placeholder="Home Phone"
    pattern="\([0-9]{3}\) [0-9]{3}-[0-9]{4}"><br>
  <label for="workPhone">Work Phone:</label>
  <input type="tel" id="workPhone" placeholder="Work Phone"
    pattern="\([0-9]{3}\) [0-9]{3}-[0-9]{4}"><br>
  <label for-"url">URL:</label>
  <input type="url" id="url" placeholder="URL"><br>
  <label for="address1">Address</label>
  <input type="text" id="address1"
    placeholder="Address (line 1)"><br>
  <label for="address2">Address</label>
  <input type="text" id="address2"
    placeholder="Address (line 2)"><br>
  <label for="city">City</label>
  <input type="text" id="city" placeholder="City"><br>
  <label for="state">State</label>
  <select id="state">
    <option>State</option>
    <option>Alabama</option>
    <option>...</option>
    <option>Washington</option>
  </select><br>
  <label for="zip">Zip Code</label>
  <input type="number" id="zip" placeholder="Zip Code"><br>
  <label for="country">Country</label>
  <select id="country">
    <option>United States</option>
    <option>...</option>
  </select><br>
  <input type="submit" value="Contact">
</form>
```

As you can see in Figure 9.1, this results in a form that is difficult to read and has little style applied to it.

NOTE

Always Include `<label>` Tags

You might notice that the HTML for the form includes `<label>` tags for each control. This is important because your pages will not be accessible if you don't label every form control. If you cannot
use the `<label>` tag, you can use attributes like `aria-label`, `aria-labelledby`, or `title` to
provide the information to screen readers. To learn more, visit the W3C website at http://
www.w3.org/TR/aria-in-html/.

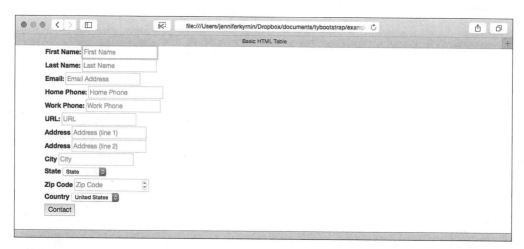

FIGURE 9.1
A basic HTML5 form.

Unlike other HTML tags we've worked with before, you need to do more than add a class to the
`<form>` tag. Some styles are applied to all the form controls, but to get better-looking forms, you
need to wrap each label and control in a `.form-group` class and add the `.form-control` class
to the controls themselves. Listing 9.2 shows how.

LISTING 9.2   Wrapping a Form Control in a Div

```
<div class="form-group">
  <label for="firstName">First Name:</label>
  <input type="text" autofocus required id="firstName"
    placeholder="First Name" class="form-control">
</div>
```

If you wrap all the form controls in a similar `<div>`, you will end up with a form that looks
like Figure 9.2.

FIGURE 9.2
A basic HTML5 form with Bootstrap classes.

As you can see, these classes give the table more space between the elements and the form controls are automatically sized to 100% of the current grid element.

Horizontal Forms

The default form in Bootstrap is vertical with the form labels directly above the controls. But Bootstrap offers two other layouts: horizontal forms and inline forms.

With horizontal forms, you can use the Bootstrap predefined grid classes to create a form where the labels are in one column and the controls are in another. You add the .form-horizontal class to your `<form>` tag, add the `.control-label`, and then add the grid classes around the labels and form controls.

Figure 9.3 shows how a horizontal form would look, and Listing 9.3 shows the HTML for the first two rows.

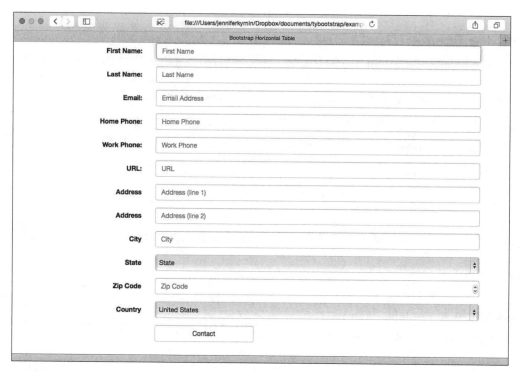

FIGURE 9.3
A horizontal form.

LISTING 9.3   First Two Rows of a Horizontal Form

```
<form action="#" class="form-horizontal">
  <div class="form-group">
    <label for="firstName" class="col-sm-3 control-label">First
    Name:</label>
    <div class="col-sm-9">
      <input type="text" autofocus required id="firstName"
      placeholder="First Name" class="form-control">
    </div>
  </div>
  <div class="form-group">
    <label for="lastName" class="col-sm-3 control-label">Last
    Name:</label>
```

```
  <div class="col-sm-9">
    <input type="text" required id="lastName"
    placeholder="Last Name" class="form-control">
  </div>
</div>
```

Inline Forms

Sometimes you want the form elements to line up next to each other rather than stack vertically. In devices that are at least 768px wide, you can use the .form-inline class on the <form> or other container element to create an inline form as in Figure 9.4 and Listing 9.4.

FIGURE 9.4
An inline form.

LISTING 9.4   An Inline Form

```
<form action="#" class="form-inline">
  <div class="form-group">
    <label for="email">Email Address</label>
    <input type="email" required id="email"
    placeholder="Email Address" class="form-control">
  </div>
  <div class="form-group">
    <label for="password">Password</label>
    <input type="password" required id="password"
    placeholder="Password" class="form-control">
  </div>
  <div class="checkbox">
    <label>
      <input type="checkbox"> Remember me
    </label>
  </div>
  <button type="submit" class="btn btn-default">Login</button>
</form>
```

NOTE

HTML5 Doesn't Require a `<form>` Tag

HTML5 allows you to put form controls anywhere on a page either inside or outside of a `<form>` tag. If you don't use the `action` attribute on the `<form>` tag, you can activate your forms using JavaScript.

As you saw in Figure 9.4, including the labels and the placeholder text in the same form can look awkward, but you can use the `.sr-only` class to hide the labels from non-screen readers.

CAUTION

Do Not Use Placeholders Alone as Labels

It can be tempting to use the `placeholder` attribute as a label for your form elements. But this can make your pages inaccessible as screen readers, and older browsers don't display placeholder text.

TRY IT YOURSELF ▼

Create an Inline Form with Hidden Labels

If you want to create a form with hidden labels, Bootstrap provides the class `.sr-only` to hide the labels from non-screen readers. This Try It Yourself takes you through how to adjust your form to hide the labels:

1. Open your Bootstrap website in a web browser.

2. Create your web form. Be sure to use `<div>` tags around the entire form group, a `<label>` tag around the label, and another `<div>` tag around the form control:

```
<div class="form-group">
  <label for="email">Email Address</label>
  <div>
    <input type="email" required id="email"
    placeholder="Email Address" class="form-control">
  </div>
</div>
```

3. Add the class `.sr-only` to the `<label>` tag:

```
<label class="sr-only" for="email">Email Address</label>
```

4. Open the form in a web browser to test that the labels are hidden. If you have access to a screen reader, you should test in that as well.

As shown in Figure 9.5, the form looks much cleaner. Listing 9.5 shows the full HTML for the form.

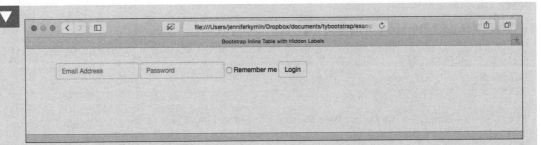

FIGURE 9.5
Inline form with hidden labels.

LISTING 9.5   HTML for an Inline Form with Hidden Labels

```
<form action="#" class="form-inline">
  <div class="form-group">
    <label for="email" class="sr-only">Email Address</label>
    <div>
    <input type="email" required id="email"
    placeholder="Email Address" class="form-control">
    </div>
  </div>
  <div class="form-group">
    <label for="password" class="sr-only">Password</label>
    <div>
    <input type="password" required id="password"
    placeholder="Password" class="form-control">
    </div>
  </div>
  <div class="checkbox">
    <label>
      <input type="checkbox"> Remember me
    </label>
  </div>
  <button type="submit" class="btn btn-default">Login</button>
</form>
```

NOTE

Check Your Form Field Widths

Bootstrap assigns a width of 100% for standard form controls. To create inline forms, the width is reset to auto. This can result in strange designs, so always test your inline forms. If you need to set a width on some of the fields, you can use the `id` attribute to set different widths on specific fields. For example:

```
#email { width: 100px; }
```

will set the width of the email form control in Listing 9.5 to 100 pixels.

Form Controls Supported by Bootstrap

HTML5 offers a lot of different form controls to collect specific data in your web forms, and Bootstrap offers support for most of them.

Basic Input Tags

The HTML5 `<input>` tag has 16 types:

- ▶ checkbox
- ▶ color
- ▶ date
- ▶ datetime
- ▶ datetime-local
- ▶ email
- ▶ month
- ▶ number
- ▶ password
- ▶ radio
- ▶ search
- ▶ tel
- ▶ text
- ▶ time
- ▶ url
- ▶ week

The HTML for the `<input>` tag looks like this in Bootstrap:

```
<input type="text" class="form-control" id="textField">
```

Change the `id` to reflect the name of the form field. You must include the `type` attribute and the `class="form-control"` for your input controls to display correctly. Use the correct type for the data you want to collect. Plus, you can use other `<input>` attributes as you need them.

CAUTION

Not All Types Change in All Browsers

The benefit of using the different specific HTML5 input types is that you can collect specific information, such as URLs or email addresses, numbers, or dates. Most modern browsers change the display of the different types to collect the data more effectively. But even if a `month` input type doesn't display a calendar in the browser, it will still display as a text field that you can validate for the data you need. You can learn more about these in my book *Sams Teach Yourself HTML5 Mobile Application Development in 24 Hours*.

Bootstrap also supports the `<textarea>` form tag. This gives you a multiline input field. This field works the same as text areas outside of Bootstrap, but you don't need the `cols` attribute because Bootstrap automatically sizes the control to 100% wide. The HTML looks like this:

```
<textarea id="textAreaField" rows="4" class="form-control">
</textarea>
```

Change the `id` to reflect the name of your text area. Add as many or as few `rows` as you need. As with all other form controls, include the `class="form-control"` so that it's styled correctly.

If you need to create the effect of a read-only form field, you can do that with the `.form-control-static` class on a `<p>` tag:

```
<p class="form-control-static">email@example.com</p>
```

Place this where you would normally place a form control. The paragraph will then have properties more like a form control, but it will be read-only.

Checkboxes and Radio Buttons

Bootstrap adds a few additional classes for checkboxes and radio buttons. The two you always use are `.checkbox` and `.radio`. Place those in container element around the form control. You can create disabled controls with the standard HTML `disabled` attribute. But if you add the Bootstrap class `.disabled` to the container element, that will display a "not allowed" cursor when the label is moused over. Listing 9.6 shows how to create checkboxes and radio buttons in Bootstrap.

LISTING 9.6   Bootstrap Checkboxes and Radio Groups

```
<div class="checkbox">
  <label>
    <input type="checkbox" value="one">
    Option one
```

```
    </label>
  </div>
  <div class="checkbox">
    <label>
      <input type="checkbox" value="two">
      Option two
    </label>
  </div>
  <div class="checkbox disabled">
    <label>
      <input type="checkbox" value="three" disabled>
      Option three - disabled
    </label>
  </div>
  <div class="radio">
    <label>
      <input type="radio" value="r-one">
      First radio button
    </label>
  </div>
  <div class="radio">
    <label>
      <input type="radio" value="r-two">
      Second radio
    </label>
  </div>
  <div class="radio disabled">
    <label>
      <input type="radio" value="r-three" disabled>
      Third radio - disabled
    </label>
  </div>
```

As you can see from Figure 9.6, when you mouse over the label of a disabled item, the cursor will change to give additional visual cues that the field is disabled.

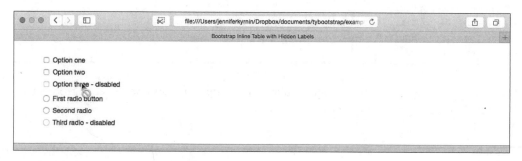

FIGURE 9.6
Checkboxes and radio buttons with a disabled field.

You can also create inline checkboxes and radio groups with the .checkbox-inline and .radio-inline classes. Use these classes in place of the .checkbox and .radio classes in the container element. These will create form fields that look like Figure 9.7.

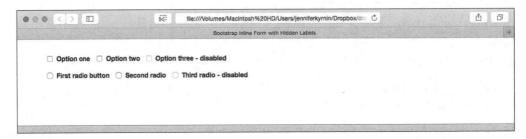

FIGURE 9.7
Inline checkboxes and radio buttons.

Sometimes it can be useful to have just a checkbox or radio button on your page without any label text at all. You can do this with Bootstrap, but best practices recommend that you include some type of label for accessibility. The easiest way is to include your label with the aria-label attribute, like this:

```
<input type="checkbox" id="soloCheckbox" value="value"
aria-label="Label for assistive devices">
```

Dropdown Menus

You can create dropdown menus with the HTML <select> and <option> tags. You create these just like you would normal HTML dropdown menus. Just make sure to use the .form-control class to have them styled correctly. Listing 9.7 shows a standard dropdown menu.

LISTING 9.7   A Standard Dropdown Menu

```
<select id="dropdown" class="form-control">
  <option>pick one</option>
  <option>one</option>
  <option>two</option>
  <option>three</option>
</select>
```

You can also use the multiple attribute to allow for multiple answers and the <optgroup> tag to group the menu options in the menu.

Change the Corners of a Dropdown Menu

Many web browsers give dropdown menus rounded corners by default. Bootstrap does not remove those rounded corners in the CSS reset (see Hour 4, "Understanding Normalize.CSS and the Basics of Bootstrap CSS"), so if you want the corners to be straight, you're going to need to adjust your custom CSS. Here's how:

1. Open your Bootstrap page in an HTML editor.

2. Add your dropdown menu to the page using the `<select>` and `<option>` tags as mentioned previously.

3. Give your menu an id of #myMenu:

   ```
   <select id="myMenu" class="form-control">
   ```

4. In the `<head>` of your document, add a link to your personal style sheet. Make sure that the style sheet comes after the Bootstrap CSS:

   ```
   <link href="css/myStyles.css" rel="stylesheet">
   ```

5. Open your style sheet in the editor.

6. Add the following CSS:

   ```
   #myMenu {
     -moz-border-radius: 0px;
     -webkit-border-radius: 0px;
     border-radius: 0px;
   }
   ```

7. Save the CSS file and the page, and test it in your web browser and other devices.

By removing the rounded corners, you can make the dropdown menu fit into designs where it might otherwise look awkward.

Sizing Form Controls

Bootstrap will let you size the width and height of form controls. To size the width of the controls, use the grid column classes such as `.col-md-*`. You learned about those in Hour 5, "Grids and How to Use Them."

Sizing the height is just as easy. Add the classes `.input-lg` and `.input-sm` to the form controls you would like to make large or small, respectively:

```
<input type="text" id="textField" class="form-control input-lg">
<input type="text" id="textField" class="form-control input-sm">
```

You can also size all the elements inside a form group by adding `.form-group-lg` or `.form-group-sm` to the form group container:

```
<div class="form-group form-group-lg">
<div class="form-group form-group-sm">
```

This will size the labels and form controls within horizontal forms.

Help Blocks

Bootstrap provides the class `.help-block` to define blocks of text that describe and inform users about the form fields. Place that class on a `<p>` or `<span>` tag that follows the form field:

```
<span id="helpfield" class="help-block">This text describes a
form field.</span>
```

You should then associate the help text with the form field it applies to with the `aria-describedby` attribute. Screen readers will then announce the help text when the user focuses or enters the form control:

```
<input type="text" id="inputWithHelpBlock" class="form-control"
aria-describedby="helpfield">
```

Make sure the `aria-describedby` attribute points to the `id` of the help block for that element. You can include as many help blocks as you need in the form.

Input Groups

Some forms and form controls can benefit from additional elements around the control. Input groups let you add fields before and after (or both) the `<input>` form control to help customers fill out the form more efficiently.

Basic Input Groups

You can create a form control that has a dollar sign ($) before the form field, and two empty decimal points (.00) after the field. Figure 9.6 shows how this would look.

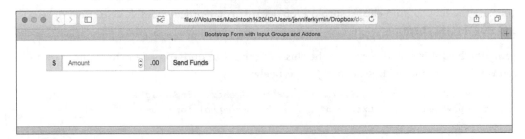

FIGURE 9.8
Inline form with input groups and add-ons.

To create this form field, you use two additional classes: `.input-group` and `.input-group-addon`. An input group is a group of items that are related to the form control, and the add-ons connect to the form control as if they were part of it. Just make sure you don't mix input groups with form groups or other grid elements. Always be sure that input groups are nested inside form groups/grid elements. Listing 9.8 shows the HTML for the form in Figure 9.8.

LISTING 9.8   Form with Input Group

```
<form action="#" class="form-inline">
  <div class="form-group">
    <div class="input-group">
      <label for="cost" class="sr-only">Amount</label>
      <div class="input-group-addon">$</div>
      <div><input type="number" required id="cost"
      placeholder="Amount" class="form-control"></div>
      <div class="input-group-addon">.00</div>
    </div>
  </div>
  <button type="submit" class="btn btn-default">Send Funds</button>
</form>
```

Sizing Input Groups

Input groups will remain a default size, but you can make them both larger and smaller to suit your needs with the classes `.input-group-lg` and `.input-group-sm`. Figure 9.9 shows the three sizes of input groups, and Listing 9.9 shows how you code them.

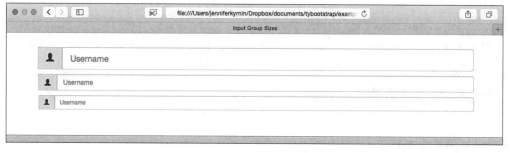

FIGURE 9.9
Different-sized input groups.

LISTING 9.9   Different-sized Input Groups

```
<p>
<div class="input-group input-group-lg">
  <label for="username" class="sr-only">Username</label>
  <span class="input-group-addon"><span
  class="glyphicon glyphicon-user"></span></span>
  <input type="text" required id="email" placeholder="Username"
  class="form-control">
</div>
</p>
<p>
<div class="input-group">
  <label for="username" class="sr-only">Username</label>
  <span class="input-group-addon"><span
  class="glyphicon glyphicon-user"></span></span>
  <input type="text" required id="email" placeholder="Username"
  class="form-control">
</div>
</p>
<p>
<div class="input-group input-group-sm">
  <label for="username" class="sr-only">Username</label>
  <span class="input-group-addon"><span
  class="glyphicon glyphicon-user"></span></span>
  <input type="text" required id="email" placeholder="Username"
  class="form-control">
</div>
</p>
```

CAUTION

Use `<span>` When You Size Input Groups

In Listing 9.8, the listing group add-ons were added on container `<div>` elements. But if you use `<div>` tags when sizing the input groups, you can end up with controls that don't match in size and look bad. If you want to test this, change the `<span class="input-group-addon">` in the `.input-group-lg` input group in Listing 9.9 and then view it in Chrome.

Fancy Add-ons

You can add many things in input groups, including checkboxes and radio buttons, dropdown menus, buttons, and combinations of them all.

To add these, you simply add HTML to the `.input-group-addon` element. Listing 9.10 shows how to add a radio button or a checkbox as an add-on. Even though you can add other HTML to the add-on area, if you do more than a short block of text or code, you should test thoroughly in browsers and smaller devices to ensure it doesn't mess up your input group.

LISTING 9.10   Radio Buttons and Checkboxes as Input Add-ons

```
<div class="row">
  <div class="col-lg-6">
    <div class="input-group">
      <span class="input-group-addon">
        <input type="checkbox" id="other" aria-label="Other"
        value="other">
      </span>
      <input type="text" id="otherText" class="form-control"
      aria-label="Other Text" placeholder="other text">
    </div>
  </div>
  <div class="col-lg-6">
    <div class="input-group">
      <span class="input-group-addon">
        <input type="radio" id="other2" aria-label="Other 2"
        value="other2">
      </span>
      <input type="text" id="otherText2" class="form-control"
      aria-label="Other Text 2" placeholder="other text">
    </div>
  </div>
</div>
```

As shown in Figure 9.10, the checkbox or radio button appears beside the input control just like the text blocks did in earlier examples.

FIGURE 9.10
Radio buttons and checkboxes as input add-ons.

You can also use buttons as input group add-ons. I discuss this in more detail in Hour 11, "Styling and Using Buttons and Button Groups."

Interactivity in Bootstrap Forms

Bootstrap provides styles for form interactivity to make the forms easier to use. These include

- ▶ **Focus**—The form field is selected in the browser.

- ▶ **Disabled**—The form field is disabled.

- ▶ **Read-only**—The form field is read-only.

- ▶ **Validation**—The form field is successful, has a warning, or has an error.

Focus State

When a form control has focus, Bootstrap removes the default `outline` styles and adds a `box-shadow`. You can use custom CSS styles to change the colors to match your styles. Listing 9.11 shows the CSS to style the focus state of an input control `<input type="text" id="formField">`.

LISTING 9.11 CSS to Restyle an Input Field Focus State

```
#formField:focus {
  border-color: #7B66E9;
  outline: 0;
  -webkit-box-shadow: inset 0 1px 1px rgba(0,0,0,.075),
    0 0 8px rgba(170,158,232,0.5);
  box-shadow: inset 0 1px 1px rgba(0,0,0,.075),
    0 0 8px rgba(170,158,232,0.75);
}
```

Disabled and Read-Only States

When you add the attribute `disabled` to a form control, it prevents users from filling in the field and changes the look of the field. You can also add the attribute to a `<fieldset>` tag to disable all the form fields inside the set:

```
<input type="text" class="form-control" id="textfld" disabled>
```

CAUTION

Disabled Fieldsets Have Some Issues

A few issues exist with using the `disabled` attribute on the `<fieldset>` tag. If you use any `<a>` tags inside the set, they will not be disabled but only given a style of `pointer-events: none`. You should use JavaScript to disable these elements. Also, Internet Explorer 11 and below don't support the `disabled` attribute on a `<fieldset>`. So you should add custom JavaScript as a fallback option.

You can make any form control read-only by adding the `readonly` attribute to the form:

```
<input type="text" class="form-control" id="textfld" readonly>
```

Validation States

You can define validation states for errors, warnings, and success using the classes `.has-error`, `.has-warning`, and `.has-success`. Figure 9.11 shows how these look; Listing 9.12 shows how the HTML looks.

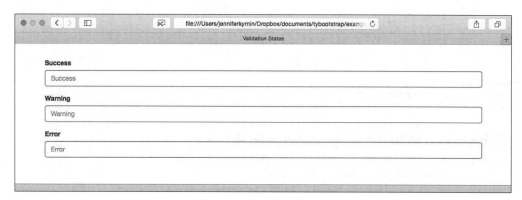

FIGURE 9.11
Validation states on form fields.

LISTING 9.12   Validation States on Form Fields

```
<form>
  <div class="form-group has-success">
    <label for="textSuccess">Success</label>
    <input type="text" class="form-control" id="textSuccess"
    placeholder="Success">
  </div>

  <div class="form-group has-warning">
    <label for="textWarning">Warning</label>
    <input type="text" class="form-control" id="textWarning"
    placeholder="Warning">
  </div>

  <div class="form-group has-error">
    <label for="textError">Error</label>
    <input type="text" class="form-control" id="textError"
    placeholder="Error">
  </div>
</form>
```

Add a feedback icon by adding the .has-feedback class to the form group and using the appropriate Glyphicon (discussed in more detail in Hour 10, "Images, Media Objects, and Glyphicons") along with the .form-control-feedback class. Figure 9.12 is a form element with a feedback icon. Listing 9.13 shows the HTML.

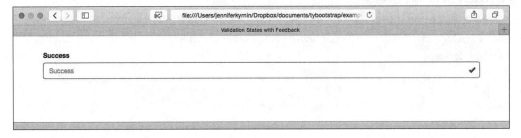

FIGURE 9.12
Validation states with feedback icon.

LISTING 9.13   **Validation States with Feedback Icon**

```
<div class="form-group has-success has-feedback">
  <label for="textSuccess">Success</label>
  <input type="text" class="form-control" id="textSuccess"
  placeholder="Success">
  <span class="glyphicon glyphicon-ok form-control-feedback"
  aria-hidden="true"></span>
</div>
```

Summary

This hour covered creating and styling web forms. You learned how to create basic forms and lay them out vertically (the default), horizontally, and inline.

Bootstrap supports a large number of form controls. This hour covered basic input tags, checkbox and radio buttons, creating dropdown menus with the <select> and <option> tags, adjusting the size of the form fields, and adding help blocks to assist users with the form.

Input groups help make the text input fields easier to use by adding additional text, icons, and other fields. This hour explained how to use input groups and how to size them to work well with your design.

Finally, you learned about the many styles that Bootstrap provides for interacting with the forms. This hour covered the focus, disabled, and read-only states as well as the three validation states for forms.

Table 9.1 shows the CSS classes Bootstrap adds for forms.

TABLE 9.1   Bootstrap Form Classes

CSS Class	Description
.checkbox	Indicates that the control is a checkbox.
.checkbox-inline	Indicates that the control should be styled as an inline checkbox.
.control-label	Indicates the element is a label for a form control.
.form-control	The element is a form control. This class should be placed on all Bootstrap form controls.
.form-control-feedback	Placed on icons to provide visual feedback of validation states on form controls.
.form-group	Indicates that the element contains a group of form controls.
.form-group-lg	Makes the form group fields larger.
.form-group-sm	Makes the form group fields smaller.
.form-horizontal	Places the form labels horizontally beside the form controls.
.has-error	Indicates that the form control has an error state.
.has-feedback	Indicates that the form control has a feedback icon.
.has-success	Indicates that the form control has a success state.
.has-warning	Indicates that the form control has a warning state.
.input-group	Indicates that the element contains an input group.
.input-group-addon	Indicates the element is an add-on to the front or back of an input group.
.input-group-lg	Makes the input group larger.
.input-group-sm	Makes the input group smaller.
.input-lg	Makes the form control larger.
.input-sm	Makes the form control smaller.
.radio	Indicates the control is a radio button.
.radio-inline	Indicates the control is an inline radio button.

Workshop

The workshop contains quiz questions to help you process what you've learned in this hour. Try to answer all the questions before you read the answers.

Q&A

Q. What is the difference between a disabled and a read-only form control?

A. A read-only form field can still get focus, and when the form is submitted, the value it has (if any) is submitted with the rest of the form. Disabled elements cannot get focus and are not submitted with the form fields. Bootstrap displays them with slightly different background colors, but otherwise they are the same.

Q. What is the difference between a static control and a read-only form control?

A. The read-only form control will still appear to be a fillable form element, while static controls will just look like text. Static controls are best used for fields that are never modified after they are created, such as usernames. This tells the reader what information is going to be automatically submitted, but there is no visual prompt to change it.

Q. Will Bootstrap form validation states show up in screen readers?

A. No. Like other contextual classes, the form validation states provide only visual feedback. You need to add `aria-describedby` attributes pointing to the validation labels so that screen readers know to read them. You can do this with help blocks that appear along with the validation states or with labels that are hidden with the `.sr-only` class.

Quiz

1. Why are labels important in Bootstrap forms?

   a. Bootstrap won't style a form control without a label.

   b. Form controls won't work without labels.

   c. Labels make the form accessible to screen readers.

   d. Labels are not required.

2. How does a horizontal form differ from a standard Bootstrap styled form?

   a. Horizontal forms have the labels displayed on the same line as the form control, while regular tables place the label directly above the form control.

   b. Horizontal forms take up 100% of the width, while regular forms take up only as much width as the control needs.

   c. Horizontal forms have wider form labels than standard forms.

   d. There is no difference. Horizontal forms are the default for Bootstrap.

3. True or False: The class `.col-md-3` grid class is used on the form control to size horizontal forms.

4. True or False: Bootstrap automatically displays a calendar on `date` and `datetime` input controls.

5. Which of these is a Bootstrap class for styling input form controls?

    a. `.color`

    b. `.url`

    c. `.radio`

    d. `.week`

6. True or False: The `multiple` attribute is a feature of Bootstrap.

7. What does the `.input-lg` class do?

    a. It makes all input tags in the input group larger.

    b. It makes all input tags in the form group larger.

    c. It makes all input tags in the form larger.

    d. It makes the input tag it is applied to larger.

8. What is an input group in Bootstrap?

    a. A group of input tags

    b. Any group of form tags

    c. A group of elements that create a standalone field with icons or text at the beginning or end (or both) of the input control

    d. Any group of tags that create form fields

9. How can you make input groups smaller?

    a. `.input-sm`

    b. `.input-group-sm`

    c. `.form-group-sm`

    d. All of the above

10. What does Bootstrap do for focused form fields?

    a. It changes the background color of the field.

    b. It changes the box-shadow on the field.

    c. It changes the font color of the field text.

    d. It doesn't change the field at all.

Quiz Answers

1. c. Labels make form controls accessible to screen readers.

2. a. Horizontal forms have the labels displayed on the same line as the form control, while regular tables place the label directly above the form control.

3. True.

4. False. Calendars are added by the browser or fallback scripts.

5. c. `.radio` is a class to help style radio buttons.

6. False. The `multiple` attribute is part of HTML, not specifically Bootstrap.

7. d. The `.input-lg` makes the input tag it is applied to larger.

8. c. An input group is a group of elements that create a standalone field with icons or text at the beginning or end (or both) of the input control.

9. b. `.input-group-sm`

10. b. It changes the box-shadow on the field.

Exercises

1. Add a form to your Bootstrap web page. Be sure to include form groups and input groups.

2. Add a dropdown menu as an add-on to an input group in your form. This works the same as adding a checkbox to an input group, except you use the `<select>` and `<option>` tags.

HOUR 10

Images, Media Objects, and Glyphicons

What You'll Learn in This Hour:

- ▶ How to add images to Bootstrap pages
- ▶ How to make images responsive
- ▶ How to change the shapes of your images
- ▶ How to add text blocks beside images with media objects
- ▶ How to add text blocks below images with thumbnails
- ▶ How to use the free Glyphicons font to add icons

Images are a useful part of any website, and Bootstrap offers several styles to help you get the most out of your images. Plus, Bootstrap offers a free font called Glyphicons that you can use as scalable icons all over your site.

Images

You can add images to your web pages using standard HTML tags. Make sure to always include the src and alt attributes (the only required attributes of the tag), and place the image where you want it on the page. Here is a standard HTML image tag:

```
<img src="myimage.gif" alt="my image">
```

CAUTION

Always Include Alt Text

It is very common to see web pages with image tags where the alt attribute, if it's included at all, is written alt="". But this is a bad habit to get into. It makes your pages much less accessible to screen readers and other assistive devices.

Responsive Images

Bootstrap helps you make your images responsive by adding the class `.img-responsive` to the image:

```
<img src="myimage.gif" alt="my image" class="img-responsive">
```

This applies the styles `max-width: 100%` and `height: auto;` to the image, so that it will scale within the parent element. Be aware that if you use an SVG image with the `.img-responsive` class, Internet Explorer 8–10 might size the image disproportionately. To fix this, add `width: 100% \9;` to your personal style sheet for that image. The `\9;` makes the property invalid, so other browsers will ignore it, while Internet Explorer 8–10 will apply the rule. Be aware that this is only a hack and might not work as expected even in Internet Explorer. Test your pages thoroughly.

You should always create your images as large as you might need for a responsive layout and then allow the CSS to resize your images down to a more manageable size. However, this means that you always need a container element with some size set on it, so that your images are not so large they scroll off the screen. If you have your page in grid elements (see Hour 5, "Grids and How to Use Them"), you can assign grid classes directly to the images to set the sizes.

Image Shapes

But there is more to images than just making them responsive. Bootstrap offers three different shapes/designs for your images. There are three classes:

- ► `.img-rounded`—The image is given rounded corners.
- ► `.img-circle`—The image is cropped into a circle.
- ► `.img-thumbnail`—Thumbnails with a narrow box surrounding them.

To get the different effects, simply add one of these classes to your images as in Listing 10.1. Figure 10.1 shows you what they look like.

LISTING 10.1 Image Shapes

```
<img src="images/dandy-header-bg.png" alt="Dandylions"
  class="img-rounded">
<img src="images/dandy-header-bg.png" alt="Dandylions"
  class="img-circle">
<img src="images/dandy-header-bg.png" alt="Dandylions"
  class="img-thumbnail">
```

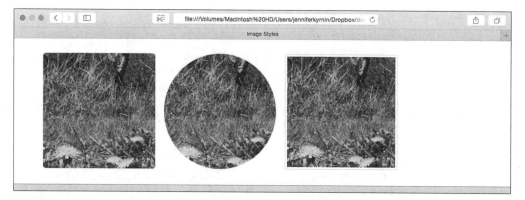

FIGURE 10.1
Image shapes.

CAUTION

There Are Some Issues with the Image Shape Classes

Internet Explorer 8 does not support rounded corners, so `.img-circle` and `.img-rounded` won't work at all and the corners on `.img-thumbnail` will be straight. If you put grid classes directly on the images to resize them (rather than on a container element like a `<div>` tag), the rounded corners on `.img-rounded` do not show up in all browsers.

Media Objects

Media objects are a way to add components that feature both textual content and an image or other media (video and audio) aligned to the right or left of the text.

NOTE

HTML Instead of Media

You aren't limited to images, video, and audio in your media objects. You can place HTML in the block to replicate pull quotes. But it is better to use the `.pull-left` or `.pull-right` classes. Learn more about the `.pull-*` classes in Hour 7, "Bootstrap Typography."

Bootstrap provides several classes for defining media objects and media groups:

▶ `.media`

▶ `.media-body`

▶ `.media-bottom`

▶ .media-heading

▶ .media-left

▶ .media-list

▶ .media-middle

▶ .media-object

▶ .media-right

The .media class defines the media group, the .media-object defines the object, and the .media-list class defines a media list. The other classes relate to how the media object is positioned on the page.

▼ TRY IT YOURSELF

Build a Media Object with Two Elements

The media object can seem confusing, but by following these steps, you'll be able to create one quickly and easily:

1. First, add a `<div>` with the class .media:

   ```
   <div class="media"></div>
   ```

2. Place a second `<div>` tag inside the .media `<div>`, and give it a class of .media-left:

   ```
   <div class="media-left"></div>
   ```

   This is the container for a media object on the left of the text.

3. Inside the .media-left `<div>` add an image with the .media-object class on it:

   ```
   <img class="media-object" src="..." alt="...">
   ```

   This is the media object. Make sure that you've included size information in your CSS for the image so that it doesn't take up the entire screen.

4. After the .media-left `<div>` add another `<div>` with the .media-body class:

   ```
   <div class="media-body"></div>
   ```

5. Inside this `<div>`, you can place a headline with the .media-heading class. This is not required:

   ```
   <h4 class="media-heading">My Media Heading</h4>
   ```

6. Then place any other text or HTML you want in the .media-body section.

7. After the `.media-body` `<div>`, you can add another `<div>` with the `.media-right` class to put a media object on the right. You build this exactly like you did the `.media-left` `<div>`:

```
<div class="media-right">
  <img class="media-object" src="..." alt="...">
</div>
```

Figure 10.2 shows what the previous Try It Yourself looks like with actual images, and Listing 10.2 shows the HTML to create it.

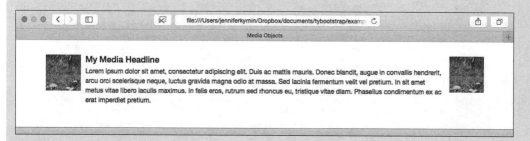

FIGURE 10.2
A media object with an image on both sides.

LISTING 10.2 A Media Object with an Image on Both Sides

```
<div class="media">
  <div class="media-left">
    <img src="images/dandy-header-bg.png" class="media-object"
      alt="dandylions">
  </div>
  <div class="media-body">
    <h4 class="media-heading">My Media Headline</h4>
    <p>Lorem ipsum dolor sit amet, consectetur adipiscing elit.
    Duis ac mattis mauris. Donec blandit, augue in convallis
    hendrerit, arcu orci scelerisque neque, luctus gravida magna
    odio at massa. Sed lacinia fermentum velit vel pretium. In sit
    amet metus vitae libero iaculis maximus. In felis eros, rutrum
    sed rhoncus eu, tristique vitae diam. Phasellus condimentum ex
    ac erat imperdiet pretium.</p>
  </div>
  <div class="media-right">
    <img src="images/dandy-header-bg.png" class="media-object"
      alt="dandylions">
  </div>
</div>
```

After you have one media object, you can add more to your page or even nest one inside another. If you have a lot of content, with smaller images, you can adjust where the media object is positioned vertically as well as horizontally. The default is top aligned, but you can also align them with the middle of the text with .media-middle or with the bottom with .media-bottom. You can see how this looks in Figure 10.3.

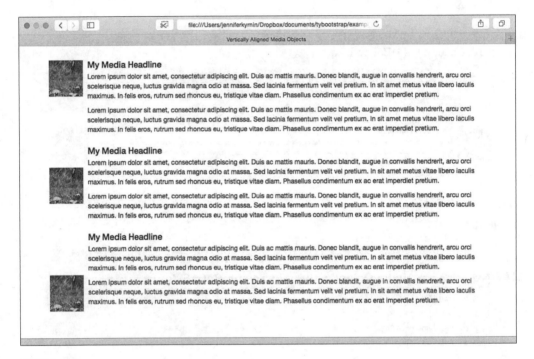

FIGURE 10.3
Vertically aligned media objects.

You can also use media objects inside a list with the .media-list class. To build a media list, you create an unordered list and apply the .media-list class to it. This removes the padding on the list and changes the list type to none, so there are no bullets or numbers. After you have defined the .media-list, put the .media class on the <li> tags and place the media objects and media bodies as you would in a non-list. Listing 10.3 shows the HTML for a simple media list.

LISTING 10.3   A Simple Media List

```
<ul class="media-list">
  <li class="media">
    <div class="media-left">
      <img src="images/dandy-header-bg.png" class="media-object"
        alt="dandylions">
```

```
    </div>
    <div class="media-body">
      <h4 class="media-heading">My Media List Headline</h4>
      <p>Lorem ipsum dolor sit amet, consectetur adipiscing elit.
      Duis ac mattis mauris. Donec blandit, augue in convallis
      hendrerit, arcu orci scelerisque neque, luctus gravida magna
      odio at massa. Sed lacinia fermentum velit vel pretium. In
      sit amet metus vitae libero iaculis maximus. In felis eros,
      rutrum sed rhoncus eu, tristique vitae diam. Phasellus
      condimentum ex ac erat imperdiet pretium.</p>
    </div>
  </li>
</ul>
```

This makes it very easy to create lists of items such as Twitter streams or product photos with descriptions using PHP or other scripts to generate the list automatically.

Thumbnails

Often when people think of images, and especially photo galleries, they think of thumbnails. These are groups of small images that are linked to a larger version of the image. You can use the Bootstrap grid system to create nicely laid out groups of thumbnails with the .thumbnail class.

To create a thumbnail listing like in Figure 10.4, first be sure you have the grid .container and .row elements; see Hour 5 for more details.

FIGURE 10.4
Thumbnails in a grid.

Then you add your image inside another container element. It could be a link (<a>) or a <div> or whatever you need, but that container has the .thumbnail class applied to it. Listing 10.4 shows how Figure 10.4 was coded.

LISTING 10.4   Thumbnails in a Grid

```
<div class="container">
  <div class="row">
    <div class="col-xs-6 col-md-2 thumbnail">
      <img src="images/thumb1.png" alt="Pets">
    </div>
    <div class="col-xs-6 col-md-2 thumbnail">
      <img src="images/thumb2.png" alt="Pets">
    </div>
    <div class="col-xs-6 col-md-2 thumbnail">
      <img src="images/thumb3.png" alt="Pets">
    </div>
    <div class="col-xs-6 col-md-2 thumbnail">
      <img src="images/thumb4.png" alt="Pets">
    </div>
    <div class="col-xs-6 col-md-2 thumbnail">
      <img src="images/thumb1.png" alt="Pets">
    </div>
    <div class="col-xs-6 col-md-2 thumbnail">
      <img src="images/thumb2.png" alt="Pets">
    </div>
  </div>
</div>
```

NOTE

Use Equal-size Thumbnails

Although you can use thumbnails of any size, the .thumbnail class works best when all the thumb-nails are the same size. Otherwise, you will get blank spots and strange stacking. If you need a presentation with different-sized images, you should check out the Masonry plugin (http://masonry.desandro.com/).

You also can add captions below your thumbnail images. To add captions to Listing 10.4, you should first separate out the grid columns <div> from the .thumbnail. For example:

```
<div class="col-xs-6 col-md-2">
  <div class="thumbnail">
    <img src="images/thumb1.png" alt="Pets">
  </div>
</div>
```

Then, inside the <div class="thumbnail"></div> you can place another container with a .caption class. Listing 10.5 shows the HTML.

LISTING 10.5 Thumbnail Image with Caption

```
<div class="col-xs-6 col-md-2">
  <div class="thumbnail">
    <img src="images/thumb1.png" alt="Pets">
    <div class="caption">
    Shasta &amp; McKinley
    </div>
  </div>
</div>
```

You can include any HTML inside the caption that you want, including headlines, buttons, and even other images or icons. Figure 10.5 shows thumbnail images with captions.

FIGURE 10.5
Thumbnails with captions.

Glyphicons

Glyphicons are a font family of 260 icons from the Glyphicon Halflings set. Normally these icons are not available for free, but the creator has made them available to Bootstrap customers for free. As a thank you, you should include a link back to Glyphicons (http://glyphicons.com/) on your website.

Figure 10.6 shows the Glyphicons included in Bootstrap 3.

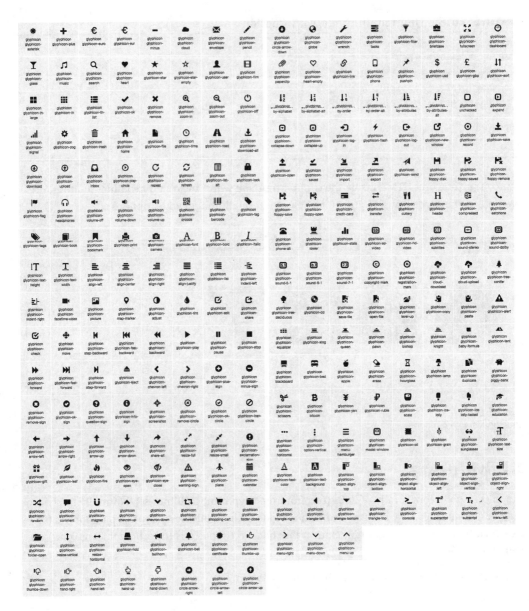

FIGURE 10.6
Glyphicons in Bootstrap.

Glyphicons are very easy to use. They require that you use the base `.glyphicon` class and the specific icon class. You place these classes on an empty `<span>` tag and place that span tag wherever you want the icon to appear. For example, Listing 10.6 shows a Glyphicon inline in some text.

LISTING 10.6   A Glyphicon Inline in Text

```
<h1>I <span class="glyphicon glyphicon-heart-empty"
style="color: red;"></span> you!</h1>
```

You will notice that there is a `style="color: red;"` attribute on the Glyphicon `<span>` tag. Because Glyphicons are a font, you can style them like you would any other font—adding color, changing the size, giving them drop shadows, and so on.

There are a few things to remember when using Glyphicons:

▶ Place all icons on an empty HTML element. The best choice is the `<span>` tag, but you can use any empty container.

▶ Always include the `.glyphicon` base class as well as the specific icon class.

▶ Do not add an icon to another component.

▶ Do not add the icon classes to an element that already has classes.

One concern with using Glyphicons is that they are solely a visual medium. This means that any customer using assistive technology won't see them and the results might be confusing. There are two possible scenarios when using Glyphicons:

▶ The icon is a decoration and doesn't add to the meaning of the content.

▶ The icon conveys some meaning.

For decorative icons, you should hide them from assistive devices with the `aria-hidden="true"` attribute. Listing 10.7 shows how.

LISTING 10.7   Icon Hidden from Assistive Devices

```
<span class="glyphicon glyphicon-ok" aria-hidden="true"></span>
```

When the icon needs to convey meaning, such as the one in Listing 10.6, you should include additional content that is hidden inside an `.sr-only` class. This will ensure that the meaning is clear to everyone who visits the page. Listing 10.8 shows how this might look, and Figure 10.7 shows what it would look like.

LISTING 10.8   Icon with Meaning

```
<h1>I <span class="glyphicon glyphicon-heart-empty"
style="color: red;" aria-hidden="true"></span>
<span class="sr-only">love</span> you!</h1>
<p class="small">
Icon courtesy <a href="http://glyphicons.com/">Glyphicons</a>.
</p>
```

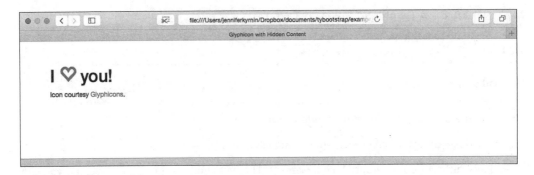

FIGURE 10.7
The word "love" is hidden from non-screen readers.

Summary

This hour covered the basics of images in Bootstrap. You learned how to make your images responsive so they fit better within multiple devices. You also learned the multiple classes Bootstrap provides to change the shape of your images.

This hour also covered how to create media objects and thumbnails to create decorative boxes and layouts with your images. Media objects let you combine images with HTML blocks in a horizontal area. Thumbnails give you a similar option, only with vertical blocks.

Finally, this hour explained how to use Glyphicons, an icon font included for free in Bootstrap.

Table 10.1 shows the classes covered in this hour; Table 10.2 lists all the Glyphicon icon classes.

TABLE 10.1   Bootstrap Classes for Images

CSS Class	Description
.caption	Descriptive text for thumbnails.
.glyphicon	The base Glyphicon class, required to use this font set. See Table 10.2 for all the specific Glyphicon classes.
.img-circle	Converts the image into a circle.
.img-responsive	Makes the image respond to the container width to flex with the size of the device.
.img-rounded	Adds rounded corners to the image.
.img-thumbnail	Indicates the image is a thumbnail or smaller copy of the image.
.media	This defines a media group containing a media object.
.media-body	The body content of a media object.

CSS Class	Description
.media-bottom	Positions the media object at the bottom of the content box.
.media-heading	The heading element of a media object.
.media-left	Positions the media object on the left.
.media-list	Defines a list of media objects.
.media-middle	Positions the media object in the vertical middle of the content box.
.media-object	The media element.
.media-right	Positions the media object on the right.
.thumbnail	Defines a thumbnail object.

TABLE 10.2 Bootstrap Glyphicon Classes

CSS Glyphicon Classes

glyphicon-adjust	glyphicon-gift	glyphicon-refresh
glyphicon-alert	glyphicon-glass	glyphicon-registration-mark
glyphicon-align-center	glyphicon-globe	
glyphicon-align-justify	glyphicon-grain	glyphicon-remove
glyphicon-align-left	glyphicon-hand-down	glyphicon-remove-circle
glyphicon-align-right	glyphicon-hand-left	glyphicon-remove-sign
glyphicon-apple	glyphicon-hand-right	glyphicon-repeat
glyphicon-arrow-down	glyphicon-hand-up	glyphicon-resize-full
glyphicon-arrow-left	glyphicon-hdd	glyphicon-resize-horizontal
glyphicon-arrow-right	glyphicon-hd-video	
glyphicon-arrow-up	glyphicon-header	glyphicon-resize-small
glyphicon-asterisk	glyphicon-headphones	glyphicon-resize-vertical
glyphicon-baby-formula	glyphicon-heart	glyphicon-retweet
glyphicon-backward	glyphicon-heart-empty	glyphicon-road
glyphicon-ban-circle	glyphicon-home	glyphicon-ruble
glyphicon-barcode	glyphicon-hourglass	glyphicon-save
glyphicon-bed	glyphicon-ice-lolly	glyphicon-save-file
glyphicon-bell	glyphicon-ice-lolly-tasted	glyphicon-saved
glyphicon-bishop		glyphicon-scale
glyphicon-bitcoin	glyphicon-import	glyphicon-scissors
glyphicon-blackboard	glyphicon-inbox	glyphicon-screenshot

CSS Glyphicon Classes

glyphicon-bold	glyphicon-indent-left	glyphicon-sd-video
glyphicon-book	glyphicon-indent-right	glyphicon-search
glyphicon-bookmark	glyphicon-info-sign	glyphicon-send
glyphicon-briefcase	glyphicon-italic	glyphicon-share
glyphicon-bullhorn	glyphicon-king	glyphicon-share-alt
glyphicon-calendar	glyphicon-knight	glyphicon-shopping-cart
glyphicon-camera	glyphicon-lamp	glyphicon-signal
glyphicon-cd	glyphicon-leaf	glyphicon-sort
glyphicon-certificate	glyphicon-level-up	glyphicon-sort-by-alphabet
glyphicon-check	glyphicon-link	
glyphicon-chevron-down	glyphicon-list	glyphicon-sort-by-alphabet-alt
glyphicon-chevron-left	glyphicon-list-alt	
glyphicon-chevron-right	glyphicon-lock	glyphicon-sort-by-attributes
glyphicon-chevron-up	glyphicon-log-in	
glyphicon-circle-arrow-down	glyphicon-log-out	glyphicon-sort-by-attributes-alt
	glyphicon-magnet	glyphicon-sort-by-order
glyphicon-circle-arrow-left	glyphicon-map-marker	glyphicon-sort-by-order-alt
	glyphicon-menu-down	
glyphicon-circle-arrow-right	glyphicon-menu-hamburger	glyphicon-sound-5-1
glyphicon-circle-arrow-up	glyphicon-menu-left	glyphicon-sound-6-1
glyphicon-cloud	glyphicon-menu-right	glyphicon-sound-7-1
glyphicon-cloud-download	glyphicon-menu-up	glyphicon-sound-dolby
glyphicon-cloud-upload	glyphicon-minus	glyphicon-sound-stereo
glyphicon-cog	glyphicon-minus-sign	glyphicon-star
glyphicon-collapse-down	glyphicon-modal-window	glyphicon-star-empty
glyphicon-collapse-up	glyphicon-move	glyphicon-stats
glyphicon-comment	glyphicon-music	glyphicon-step-backward
glyphicon-compressed	glyphicon-new-window	glyphicon-step-forward
glyphicon-console	glyphicon-object-align-bottom	glyphicon-stop
glyphicon-copy		glyphicon-subscript
glyphicon-copyright-mark	glyphicon-object-align-horizontal	glyphicon-subtitles
glyphicon-credit-card		glyphicon-sunglasses
glyphicon-cutlery	glyphicon-object-align-left	glyphicon-superscript

CSS Glyphicon Classes

glyphicon-dashboard

glyphicon-download

glyphicon-download-alt

glyphicon-duplicate

glyphicon-earphone

glyphicon-edit

glyphicon-education

glyphicon-eject

glyphicon-envelope

glyphicon-equalizer

glyphicon-erase

glyphicon-eur

glyphicon-euro

glyphicon-exclamation-sign

glyphicon-expand

glyphicon-export

glyphicon-eye-close

glyphicon-eye-open

glyphicon-facetime-video

glyphicon-fast-backward

glyphicon-fast-forward

glyphicon-file

glyphicon-film

glyphicon-filter

glyphicon-fire

glyphicon-flag

glyphicon-flash

glyphicon-floppy-disk

glyphicon-floppy-open

glyphicon-floppy-remove

glyphicon-floppy-save

glyphicon-floppy-saved

glyphicon-object-align-right

glyphicon-object-align-top

glyphicon-object-align-vertical

glyphicon-off

glyphicon-oil

glyphicon-ok

glyphicon-ok-circle

glyphicon-ok-sign

glyphicon-open

glyphicon-open-file

glyphicon-option-horizontal

glyphicon-option-vertical

glyphicon-paperclip

glyphicon-paste

glyphicon-pause

glyphicon-pawn

glyphicon-pencil

glyphicon-phone

glyphicon-phone-alt

glyphicon-picture

glyphicon-piggy-bank

glyphicon-plane

glyphicon-play

glyphicon-play-circle

glyphicon-plus

glyphicon-plus-sign

glyphicon-print

glyphicon-pushpin

glyphicon-qrcode

glyphicon-queen

glyphicon-tag

glyphicon-tags

glyphicon-tasks

glyphicon-tent

glyphicon-text-background

glyphicon-text-color

glyphicon-text-height

glyphicon-text-size

glyphicon-text-width

glyphicon-th

glyphicon-th-large

glyphicon-th-list

glyphicon-thumbs-down

glyphicon-thumbs-up

glyphicon-time

glyphicon-tint

glyphicon-tower

glyphicon-transfer

glyphicon-trash

glyphicon-tree-conifer

glyphicon-tree-deciduous

glyphicon-triangle-bottom

glyphicon-triangle-left

glyphicon-triangle-right

glyphicon-triangle-top

glyphicon-unchecked

glyphicon-upload

glyphicon-usd

glyphicon-user

glyphicon-volume-down

glyphicon-volume-off

glyphicon-volume-up

glyphicon-warning-sign

CSS Glyphicon Classes

glyphicon-folder-close	glyphicon-question-sign	glyphicon-wrench
glyphicon-folder-open	glyphicon-random	glyphicon-yen
glyphicon-font	glyphicon-record	glyphicon-zoom-in
glyphicon-forward		glyphicon-zoom-out
glyphicon-fullscreen		
glyphicon-gbp		

Workshop

The workshop contains quiz questions to help you process what you've learned in this hour. Try to answer all the questions before you read the answers.

Q&A

Q. It seems like adding the one class `.img-responsive` to an image isn't enough to make it responsive. How does that work?

A. If you've read my book *Sams Teach Yourself Responsive Web Design in 24 Hours*, you know that using flexible-width images is an easy way to get your images to flex with the browser width. The `.img-responsive` class relies on the built-in layout elements to limit the size of the image. There are many other things you can do to speed up images and make them "Retina ready," but they are beyond the scope of Bootstrap.

Q. Are there other shapes I can convert my images to with Bootstrap?

A. Bootstrap offers the three image shapes: circle, rounded, and thumbnail. But you can use CSS to adjust your images even more.

Quiz

1. Which of these attributes is required on an image tag?

   a. `alt`

   b. `class`

   c. `src`

   d. A and C

   e. All of the above

2. What does the .img-responsive class do to images?

   a. Makes the width 100%

   b. Makes the maximum width 100%

   c. Makes the height automatic

   d. B and C

   e. All of the above

3. How can you turn your image into a circle with Bootstrap?

   a. .circle

   b. .img-circle

   c. .img-rounded

   d. You can't.

4. True or False: The image shape classes work in all browsers.

5. True or False: HTML is valid inside media objects.

6. Where should you put the HTML for a .media-left element?

   a. This should come before the media body.

   b. This should come after the media body.

   c. Include it in the media body.

   d. It doesn't matter where you put it.

7. What's the difference between a .media element and a .media-list element?

   a. The .media element defines a media group and can contain a .media-list.

   b. The .media-list element defines a media list that can contain a .media element.

   c. They are both container classes for media objects, but a .media-list is placed on the <ul> or <ol> list tags.

   d. There is no difference.

8. How do you size thumbnail images in Bootstrap?

   a. The .thumbnail class sets images to 150×150 pixels.

   b. The .img-thumbnail class sets images to 50% of the container width.

   c. You use grid classes on the images.

   d. You don't; you should use the width and height attributes.

9. What type of content can go in a `.caption` element?

   a. Any HTML

   b. Headline text

   c. Caption text

   d. Images

10. Which of the following is the correct method to add an icon to a Bootstrap page?

    a. `<div class="col-xs-2 glyphicon glyphicon-bed">Column 1</div>`

    b. `<span class="glyphicon glyphicon-bed"></span>`

    c. `<span class="glyphicon-bed"></span>`

    d. `<button class="button glyphicon">button</button>`

Quiz Answers

1. d. The `<img>` tag requires the `alt` and `src` attributes.

2. d. The `.img-responsive` class sets the image to `max-width: 100%;` and `height: auto;`.

3. b. The `.img-circle` class displays the image in a circle.

4. False. The image shapes don't work in older browsers like Internet Explorer 8.

5. True. You typically use media like images, videos, or audio, but other HTML elements are valid as well.

6. a. On pages that are left-to-right (English and similar languages), the first element in the HTML is on the left, and that is where you should place the `.media-left` element.

7. c. They are both container elements for media objects, but the `.media-list` class goes on list tags.

8. c. Use the Bootstrap grid classes with the `.thumbnail` class to create thumbnail grids and resize the images.

9. a. You can use the `.thumbnail` and `.caption` classes to contain any HTML to create a vertical thumbnail box.

10. b. `<span class="glyphicon glyphicon-bed"></span>`

Exercises

1. Add an image in a shape to your Bootstrap web page.

2. Find an icon that would add either interest (decoration) or information (meaning) to your page, and add it to the page.

Styling and Using Buttons and Button Groups

What You'll Learn in This Hour:

▶ How to style buttons in Bootstrap

▶ How to resize and color buttons

▶ How to create a button group

▶ How to style button groups with color, size, and display

When you're creating interactive websites, you need a tool that indicates to the customer that clicks will cause an action. Buttons are an easily recognized interface, and Bootstrap makes it simple to add them to your web pages.

This hour you learn how to create basic buttons with Bootstrap and adjust how they look on the page. You also learn how to create groups of buttons and even dropdown menus with your buttons.

Basic Buttons

Buttons in Bootstrap look exactly like you expect. Figure 11.1 shows what a basic button looks like.

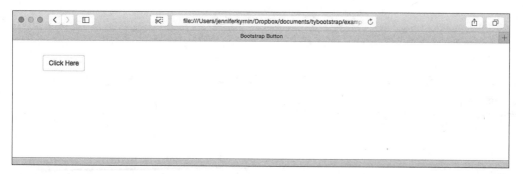

FIGURE 11.1
A standard Bootstrap button.

Bootstrap buttons are very plain by default. They have slightly rounded corners, a white background, and a thin gray border. When you mouse over them, they turn light gray, and when they are actively clicked, Bootstrap adds an inset gray shadow on the button.

Button Tags

To add a button to your page, you first add the class .btn and then the .btn-default class to create a default button. You can create buttons on three HTML tags:

- `<button>`
- `<a>`
- `<input>`

The `<input>` tag can be either a type="button" or type="submit". Add the classes to these HTML elements, as in Listing 11.1.

LISTING 11.1 Four Button Tags

```
<a class="btn btn-default">Link</a>
<button class="btn btn-default" type="submit">Button</button>
<input type="button" class="btn btn-default" value="Input">
<input type="submit" class="btn btn-default" value="Submit">
```

As shown in Figure 11.2, these tags create buttons that look the same. So you can use the HTML tags that work best on your site.

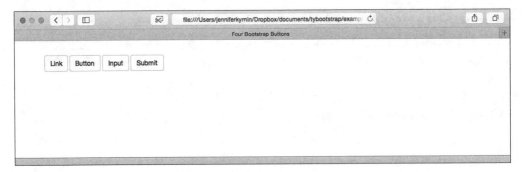

FIGURE 11.2
Four Bootstrap buttons using different HTML tags.

Button Best Practices

With Bootstrap, you can create buttons using any of the four listed tags, but there are some best practices you should follow if you can. You can use the button classes only on `<button>` elements in the navigation (see Hour 12, "Creating Navigation Systems with Bootstrap"). If you create a link as a button, you should indicate the role with the `role="button"` attribute. Also, for the widest possible future support, you should always use the `<button>` element as your first choice for buttons.

Button Classes and Sizes

Like with other Bootstrap elements, you can change the color and size of your buttons with different secondary classes. Seven classes are available:

- `btn-default`—The standard button
- `btn-primary`—The primary action in a set of buttons
- `btn-success`—Indicates a successful or positive action
- `btn-info`—Contextual button for informational alert buttons
- `btn-warning`—Indicates that caution should be taken with this action
- `btn-danger`—Indicates a dangerous or negative action
- `btn-link`—Makes the button look like a link while still behaving like a button

Convey Meaning in Other Ways

When you use the contextual classes to convey meaning, you risk that some of your customers won't access that meaning because they don't see color or they use audio browsers. Be sure that the content is presented in some other way, beyond the color of the button, to make it accessible. One way is to use the `.sr-only` class to hide content from non-screen readers.

Listing 11.2 shows you how to add these button types to your HTML, and Figure 11.3 shows how they would look.

LISTING 11.2   Seven Button Types in Bootstrap

```
<button class="btn btn-default">Default</button>
<button class="btn btn-primary">Primary</button>
<button class="btn btn-success">Success</button>
<button class="btn btn-info">Info</button>
```

```
<button class="btn btn-warning">Warning</button>
<button class="btn btn-danger">Danger</button>
<button class="btn btn-link">Link</button>
```

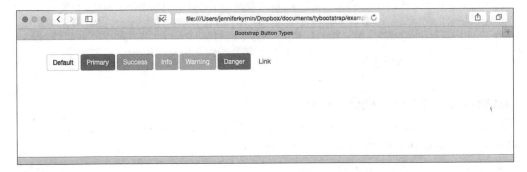

FIGURE 11.3
Seven button types in Bootstrap.

There are four sizes of buttons in Bootstrap. The default is a medium size button, but you can also use classes to create a large, small, or extra small button:

▶ `.btn-lg`

▶ `.btn-sm`

▶ `.btn-xs`

You can combine the size classes with the contextual classes to create a large warning button or an extra small primary one. You also can create buttons that span the full width of the parent element with the `.btn-block` class. Figure 11.4 shows how these size classes affect the way the buttons look. You can see the HTML for it in Listing 11.3.

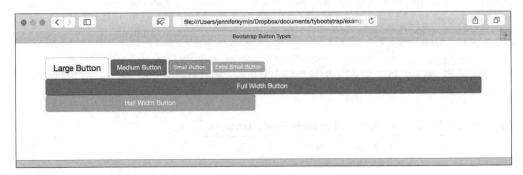

FIGURE 11.4
Change the button sizes.

LISTING 11.3   Change the Button Sizes

```
<div class="container">
  <button class="btn btn-default btn-lg">Large Button</button>
  <button class="btn btn-primary">Medium Button</button>
  <button class="btn btn-success btn-sm">Small Button</button>
  <button class="btn btn-warning btn-xs">Extra Small
  Button</button>
  <button class="btn btn-danger btn-block">Full Width
  Button</button>
  <div class="row">
    <div class="col-md-6">
        <button class="btn btn-info btn-block">Half Width
        Button</button>
    </div>
  </div>
</div>
```

The easiest way to change the size of a block button is to put it inside a column element on your grid. In Listing 11.3, I placed the button in an element six columns wide so it would span half the width of the screen.

Button States

Buttons have states, just like links. They can be active so they appear pressed or disabled so they can't be clicked on. Bootstrap adds some additional styles to these states so that the buttons appear more interactive.

Active buttons will appear pressed by getting a darker background, a darker border, and an inset shadow when they are clicked on. Figure 11.5 shows buttons in an active state.

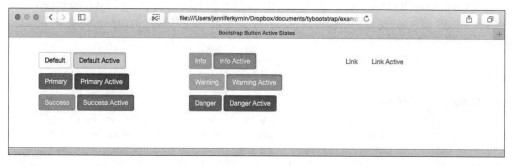

FIGURE 11.5
Active Bootstrap buttons.

Disabled buttons are not clickable, and they are given styles to fade them into the background of the page. You can make any button disabled with the `disabled` attribute. For example:

```
<button class="btn btn-default" disabled>Disabled Button</button>
```

Figure 11.6 shows how Bootstrap displays disabled buttons.

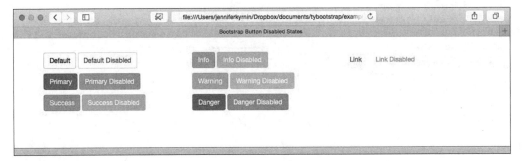

FIGURE 11.6
Disabled Bootstrap buttons.

CAUTION

Don't Use the `disabled` Attribute with `<a>` Tags

The `disabled` attribute is not a valid attribute of the `<a>` tag and can cause problems if you use it there. Most modern browsers will display a disabled button written with an anchor (`<a>`) tag correctly, but it may still be clickable, which defeats the purpose. Instead, use the `.disabled` class on the tag and disable it that way.

You use the classes `.active` and `.disabled` to force these styles on buttons that might otherwise display differently. Be aware that buttons created with an anchor (`<a>`) tag might behave slightly differently when set to `.active` or `.disabled`. Be sure to test in your devices if you are using that tag for your buttons.

NOTE

Why Disable Buttons?

It can seem odd to have the ability to disable buttons on web pages. After all, wouldn't that frustrate customers if they can't click on a button? But disabling elements on your web pages can make the forms more usable. For instance, if the form has a large number of required fields, you can set the submit button to be disabled until all the required fields are filled in. This helps customers get the form submitted accurately on the first try. You also might disable a button on a "Terms and Conditions" field, so that the "Next" button is not active until the reader has scrolled through the entire terms. You should think of the disabled attribute as a way to make forms easier to fill in and avoid using it just to annoy your customers.

Button Groups

You can group buttons together on a line as a group. This connects the buttons together and makes them a more cohesive whole. You can create horizontal, vertical, and toolbar button groups.

To create a button group, you surround your buttons with a container element such as `<div>` with the class `.btn-group`. Then be sure to include the appropriate `role` attribute: `toolbar` for toolbars and `group` for button groups. This will ensure that screen readers know that the buttons are grouped together. Finally, to be fully accessible, you should include a label for the screen readers to read, such as with the `aria-label` attribute or by including text in the buttons and hiding it with the `.sr-only` class. Listing 11.4 shows the HTML for a simple button group.

LISTING 11.4 A Simple Button Group

```
<div class="btn-group" role="group" aria-label="Button Group">
  <button type="button" class="btn btn-default">Fast</button>
  <button type="button" class="btn btn-default">Slow</button>
  <button type="button" class="btn btn-default">Stop</button>
</div>
```

Horizontal Button Groups

Horizontal button groups are the default layout for button groups.

Just like with standalone buttons, you can resize your button groups with three classes:

▶ `.btn-group-lg`

▶ `.btn-group-sm`

▶ `.btn-group-xs`

These make the button groups larger and smaller, as in Figure 11.7.

FIGURE 11.7
Different size button groups.

Another class you can use on button groups is the `.btn-group-justified` class. This will stretch the buttons in the group to span the entire width of the container.

If you're using `<a>` tags for your buttons, just add the class to the button group container, as in Listing 11.5. Figure 11.8 shows what it will look like.

LISTING 11.5   Justified Button Group

```
<div class="btn-group btn-group-justified" role="group"
    aria-label="Button Group">
  <a href="#" class="btn btn-default">Fast</a>
  <a href="#" class="btn btn-default">Slow</a>
  <a href="#" class="btn btn-default">Stop</a>
</div>
```

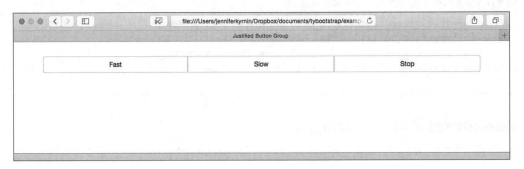

FIGURE 11.8
A justified button group.

It's a little trickier with `<button>` tags. Most browsers don't properly apply the Bootstrap CSS for justification correctly. So, to get it to work, you should surround each button with another `.btn-group` element, as in Listing 11.6.

LISTING 11.6   Justified Button Group with `<button>` Elements

```
<div class="btn-group btn-group-justified" role="group"
    aria-label="Button Group">
  <div class="btn-group">
    <button type="button" class="btn btn-default">Fast</button>
  </div>
  <div class="btn-group">
    <button type="button" class="btn btn-default">Slow</button>
  </div>
  <div class="btn-group">
    <button type="button" class="btn btn-default">Stop</button>
  </div>
</div>
```

Vertical Button Groups

If you want your button group to be stacked vertically, use the class `.btn-group-vertical`. By adding this class to the HTML in Listing 11.4, you will get a button group like in Figure 11.9.

FIGURE 11.9
A vertical button group.

Button Toolbars

You make toolbars by combining multiple button groups. This gives you the ability to style more complex elements. Listing 11.7 shows how you might write the HTML for a button toolbar with three button groups. Figure 11.10 shows what it would look like in the default Bootstrap styles with a couple of fancier buttons.

LISTING 11.7   A Button Toolbar

```
<div class="btn-toolbar">
  <div class="btn-group" role="group" aria-label="Button Group">
    <button type="button" class="btn btn-primary">1</button>
    <button type="button" class="btn btn-default">2</button>
    <button type="button" class="btn btn-default">3</button>
  </div>
  <div class="btn-group" role="group" aria-label="Button Group">
    <button type="button" class="btn btn-default">4</button>
    <button type="button" class="btn btn-default">5</button>
    <button type="button" class="btn btn-default">6</button>
    <button type="button" class="btn btn-default">7</button>
  </div>
  <div class="btn-group" role="group" aria-label="Button Group">
    <button type="button" class="btn btn-default">8</button>
    <button type="button" class="btn btn-default">9</button>
    <button type="button" class="btn btn-danger">10</button>
  </div>
</div>
```

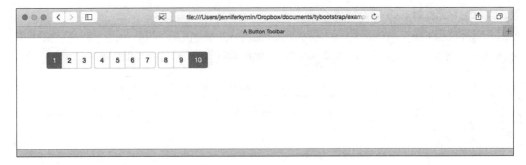

FIGURE 11.10
A button toolbar.

As you can see from Figure 11.10, you can always adjust the buttons to use the different button types.

Button JavaScript

Bootstrap includes a script to let you do more with your buttons: button.js. You can use this script to control the state of your buttons, toggle the buttons on and off, and change your checkbox and radio input fields into buttons. You will learn how to use these in Hour 18, "Transitions, Buttons, Alerts, and Progress Bars."

Summary

This hour covered how to use buttons and button groups in Bootstrap. Bootstrap provides classes to convert your <button>, <a>, and <input> tags into buttons that respond to customer interaction. You also learned how to combine multiple buttons into groups and toolbars.

You learned about the classes that create buttons and button groups and change the size and color of those buttons and groups. You also learned the classes to convert the button groups to vertical button groups and toolbars. Table 11.1 shows the classes covered in this hour.

TABLE 11.1 Bootstrap Classes for Buttons and Button Groups

CSS Class	Description
.active	Puts the element in an active state, even when it's not active.
.btn	Turns the element into a button.
.btn-block	The button is a block button, the same width as the parent element.
.btn-danger	Indicates that the button is dangerous or would have a dangerous result.

CSS Class	Description
`.btn-default`	The default button styling.
`.btn-group`	Indicates the element encloses a group of buttons.
`.btn-group-lg`	Displays the button group as large.
`.btn-group-sm`	Displays the button group as small.
`.btn-group-xs`	Displays the button group as extra small.
`.btn-info`	Indicates the button is informative or provides information.
`.btn-lg`	Displays the button as large.
`.btn-link`	Changes the button to display as a link to lessen the priority or importance of the button.
`.btn-primary`	Indicates the button provides important information.
`.btn-sm`	Displays the button as small.
`.btn-success`	Indicates that the button provides a successful or positive outcome.
`.btn-toolbar`	Indicates that the element contains several button groups that will be displayed as a toolbar.
`.btn-warning`	Indicates that the button may have a negative outcome.
`.btn-xs`	Displays the button as extra small.
`.disabled`	Styles the element in a disabled state, even if it is not disabled.

Workshop

The workshop contains quiz questions to help you process what you've learned in this hour. Try to answer all the questions before you read the answers.

Q&A

Q. How do I make the buttons do something?

A. Bootstrap is a framework for providing the look and feel of a website and a little interactivity. When you click on a button on a Bootstrap site, you need to have some type of script (either JavaScript, PHP, or something else) that does something. Hour 18 covers the basic actions that Bootstrap provides, but if you want it to do more, you should read a book like *Sams Teach Yourself jQuery and JavaScript in 24 Hours* by Brad Dayley or *Sams Teach Yourself PHP, MySQL, and Apache All in One* by Julie Meloni.

Q. I prefer to use the `<a>` tag for my buttons; does this really matter?

A. You can use the `<a>`, `<button>`, or `<input>` tags for buttons. Firefox versions below 30 have a rendering bug that might make your `<input>` buttons a slightly different height from other buttons. And other browsers might have similar differences. It's better to pick one HTML element and use that consistently for your buttons. Bootstrap recommends `<button>`, but you can use any of them.

Q. **Can I turn checkboxes and radio buttons into buttons?**

A. To make a checkbox or radio input button into a Bootstrap button, you need to do a few things. You need to use the `button.js` script, add `data-toggle="buttons"` to the button group container, and then add the `.btn` and `.btn-type` classes to the `<label>` around the input field. All this is covered in detail in Hour 18.

Quiz

1. Which of the following is a valid tag for creating a button?

    a. `<a>`

    b. `<button>`

    c. `<input>`

    d. All of the above

2. Which of the following is the best tag to use for creating buttons?

    a. `<a>`

    b. `<button>`

    c. `<input>`

    d. All of the above

3. Which class is required to create a Bootstrap button?

    a. `.btn`

    b. `.btn-default`

    c. `.btn-link`

    d. `.button`

4. What does the `.btn-link` class do?

    a. It turns a link tag (`<a>`) into a button.

    b. It creates a link that looks like a button but acts like a link.

    c. It creates a button that looks like a link.

    d. It is not a valid Bootstrap class.

5. Which of the following is not a secondary class for buttons?

    a. `.btn-correct`

    b. `.btn-default`

    c. `.btn-primary`

    d. `.btn-success`

6. Which of these classes will resize a button?

    a. `.btn-lg`

    b. `.btn-md`

    c. `.btn-small`

    d. All of the above

7. How do you disable a `<button>` element?

    a. Add the `.disabled` class.

    b. Add the `disabled` attribute.

    c. Add both the `.disabled` class and the `disabled` attribute.

    d. You cannot disable a `<button>` element.

8. When should you use the state classes `.active` and `.disabled`?

    a. Use them whenever the state changes on your buttons.

    b. Use them when you need the state to be explicitly displayed.

    c. Use them only on `<a>` tag buttons.

    d. Use them whenever you want to.

9. What is the class (or classes) to define a button group?

    a. `.btn` and `.group`

    b. `.btngroup`

    c. `.btn-group`

    d. `.group`

10. How do you ensure a button group is accessible?

    a. Use the `.btn-group` or `.btn-toolbar` classes to make it accessible.

    b. Add the `<role>` tag to the button group.

    c. Use an ARIA label.

    d. Use the `role` attribute to define it as a `toolbar` or a `group`.

Quiz Answers

1. d. All of the above.

2. b. `<button>` is the recommended best practice.

3. a. `.btn` must be on all buttons.

4. c. It creates a button that looks like a link.

5. a. The `.btn-correct` class is not a Bootstrap button secondary class.

6. a. `.btn-lg` will make the buttons larger than the default.

7. b. Add the `disabled` attribute.

8. b. You should use the state classes when you want the state to be explicitly displayed.

9. c. `.btn-group`

10. d. Use the `role` attribute to define it as a `toolbar` or a `group`.

Exercises

1. Open your Bootstrap page and add a button to the page.

2. Create a group of buttons in a button group.

Creating Navigation Systems with Bootstrap

What You'll Learn in This Hour:

▶ How to build standard navigation elements

▶ How to create dropdown and dropup menus

▶ How to add site navigation with navbars

▶ How to build breadcrumbs and pagination

▶ How to create list groups

Navigation is an important part of any website. It's how visitors get around and how the site owners position the content so it can be found.

Bootstrap offers several features for navigation. This hour you will learn how to create standard navigation and navbars, dropdown menus, breadcrumbs, and list groups, and even add pagination to your web pages.

Standard Navigation Elements

Navigation systems in Bootstrap are called *navs*. You use the base .nav class on the container element to create a nav. In HTML5 you should start with the <nav> element to define the navigation. This element can define navigation for the entire site, the page, or even just the section it's in. Include a role="navigation" attribute on the <nav> element so your navigation is accessible. Your navigation will be in a list element such as <ul>, and each item will be an <li> element with a link (<a>).

The two types of standard navigation are

▶ .nav-pills—Creates a small button for each navigation item

▶ .nav-tabs—Turns the navigation elements into simple tabs

▼ TRY IT YOURSELF

Build a Simple Navigation in Bootstrap

Bootstrap lets you easily convert a standard HTML5 navigation into a Bootstrap navigation. This Try It Yourself takes you through the steps to create a simple navigation on your Bootstrap page:

1. Open your Bootstrap page in an HTML editor.

2. Add the following HTML:

```
<nav role="navigation">
  <ul>
    <li role="presentation"><a href="#">Home</a></li>
    <li role="presentation"><a href="#">About</a></li>
    <li role="presentation"><a href="#">Articles</a></li>
    <li role="presentation"><a href="#">Support</a></li>
  </ul>
</nav>
```

3. Add the .nav class to the <ul> tag.

4. Add the .nav-pills class to the <ul> tag.

5. Add the .active class to the first <li> tag.

As shown in Figure 12.1, this creates a group of navigation links with the active button in blue. You can change the navigation to .nav-tabs to create a tabbed navigation. Listing 12.1 shows the HTML for tabs.

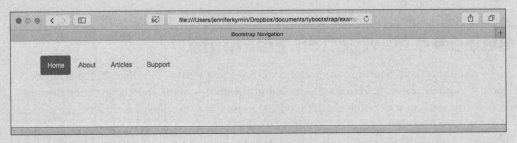

FIGURE 12.1
Navigation pills with an active button.

LISTING 12.1   HTML for Tabbed Navigation

```
<nav role="navigation">
  <ul class="nav nav-tabs">
    <li role="presentation" class="active"><a
        href="#">Home</a></li>
    <li role="presentation"><a href="#">About</a></li>
```

```
    <li role="presentation"><a href="#">Articles</a></li>
    <li role="presentation"><a href="#">Support</a></li>
  </ul>
</nav>
```

If you use the `.nav-pills` class, you can create a vertical navigation with the added class `.nav-stacked`. This will convert the navigation elements to the width of the container. You can create a standard two-column layout as in Figure 12.2 by putting the navigation in one grid column and the main content in another.

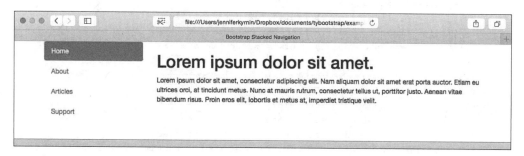

FIGURE 12.2
Two-column layout with stacked pills navigation.

As you can see in Listing 12.2, the column classes are placed right on the `<nav>` and `<article>` elements.

LISTING 12.2   Two-Column Layout with Stacked Pills Navigation

```
<!DOCTYPE html>
<html lang="en">
  <head>
    <meta charset="utf-8">
    <meta http-equiv="X-UA-Compatible" content="IE=edge">
    <meta name="viewport"
          content="width=device-width, initial-scale=1">
    <title>Bootstrap Stacked Navigation</title>

    <!-- Bootstrap -->
    <link href="css/bootstrap.min.css" rel="stylesheet">
    <!-- HTML5 shim and Respond.js for IE8 support of HTML5
    elements and media queries -->
    <!-- WARNING: Respond.js doesn't work if you view the page
    via file:// -->
    <!--[if lt IE 9]>
```

```
      <script
src="https://oss.maxcdn.com/html5shiv/3.7.2/html5shiv.min.js">
      </script>
      <script
src="https://oss.maxcdn.com/respond/1.4.2/respond.min.js"></script>
    <![endif]-->
  </head>
  <body>
  <div class="container">
    <div class="row">
    <nav role="navigation" class="col-sm-3">
      <ul class="nav nav-pills nav-stacked">
        <li role="presentation" class="active">
          <a href="#">Home</a></li>
        <li role="presentation"><a href="#">About</a></li>
        <li role="presentation"><a href="#">Articles</a></li>
        <li role="presentation"><a href="#">Support</a></li>
      </ul>
    </nav>
    <article class="col-sm-9">
      <h1>Lorem ipsum dolor sit amet.</h1>
      <p>Lorem ipsum dolor sit amet, consectetur adipiscing elit.
      Nam aliquam dolor sit amet erat porta auctor. Etiam eu
      ultrices orci, at tincidunt metus. Nunc at mauris rutrum,
      consectetur tellus ut, porttitor justo. Aenean vitae
      bibendum risus. Proin eros elit, lobortis et metus at,
      imperdiet tristique velit.</p>
    </article>
    </div>
  </div>
  </body>
</html>
```

You can make both tabs and pills the equal width with the `.nav-justified` class. This will resize the navigation links in browsers wider than 768px. Smaller than that, and the navigation links will be stacked. You can see how this looks with both pills and tabs in Figure 12.3.

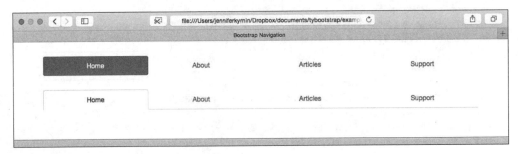

FIGURE 12.3
Justified navigation.

If you want to disable sections of your navigation, just add the `.disabled` class to the `<li>` items. For example:

```
<li role="presentation" class="disabled"><a href="#">Games</a></li>
```

This will make the links gray and remove any hover effects. Keep in mind that the `<a>` tag will still function. Use JavaScript to disable the links completely.

Dropdowns

Dropdown menus are a great way to add some simple navigation to your pages. You add a container with the `.dropdown` class surrounding a button with the `.dropdown-toggle` class. (Learn more about buttons in Hour 11, "Styling and Using Buttons and Button Groups.") Then you add the menu with the `.dropdown-menu` class. Listing 12.3 shows how it might look.

LISTING 12.3   A Dropdown Menu

```
<div class="dropdown">
  <button class="btn btn-default dropdown-toggle" data-toggle="dropdown">
    Choose One...
    <span class="caret"></span>
  </button>
  <ul class="dropdown-menu">
    <li><a href="#">Item 1</a></li>
    <li><a href="#">Item 2</a></li>
    <li><a href="#">Item 3</a></li>
    <li><a href="#">Item 4</a></li>
  </ul>
</div>
```

After you have your dropdown menu, you need to be sure to include links to jQuery and the Bootstrap JavaScript at the bottom of your HTML document. For example:

```
<script src="http://code.jquery.com/jquery-latest.js"></script>
<script src="js/bootstrap.min.js"></script>
```

Here are a few other classes you can add to your dropdown menus:

- ▶ `.dropdown-menu-left`—Forces the menu to display left aligned on the container (the default)

- ▶ `.dropdown-menu-right`—Forces the menu to display right aligned on the container

- ▶ `.dropdown-header`—Creates a header menu item

- ▶ `.divider`—Creates a divider inside the menu

- ▶ `.disabled`—Disables the menu item

Figure 12.4 shows how the `.dropdown-header`, `.divider`, and `.disabled` classes look on menu items.

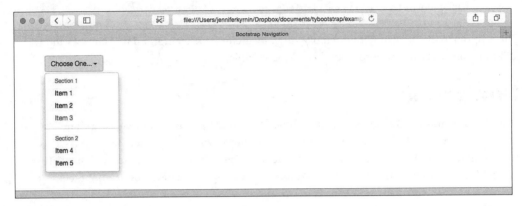

FIGURE 12.4
Dropdown menu with special classes.

CAUTION

Only One Level of Dropdown Menu

One thing that Bootstrap 3 doesn't do right out of the box is help you create multilevel dropdown menus. One level is all you get. If you need a more complex navigation structure than just the one dropdown, you need to add custom CSS and JavaScript. You can find many articles on the Web to help. This Fiddle (http://jsfiddle.net/chirayu45/YXkUT/16/) is a good example.

Split Dropdowns

Split button dropdowns are dropdown menus created on a button group that make it appear to be one single button with a dropdown menu on one side.

To create this, you need to use a button group with two buttons: the action button with text on it and a dropdown button that just has the `.caret` icon class on it to trigger the dropdown menu.

▼ TRY IT YOURSELF

Create a Split Button Dropdown Menu

If you read Hour 11, you already know how to create button groups, and that is all a split button is—a button group with a button and a dropdown. In this Try It Yourself, you will see how easily you can use a dropdown menu in the button group to create a split button:

1. Open your Bootstrap page in a web editor.

2. Create a button group with an action button:

```
<div class="btn-group">
  <button class="btn btn-primary">Choose One...</button>
</div>
```

3. Add another button below the first, and make this one your dropdown toggle:

```
<button class="btn btn-primary dropdown-toggle"
        data-toggle="dropdown">
  <span class="caret"></span>
  <span class="sr-only">Trigger Dropdown Menu</span>
</button>
```

4. Create a list with your dropdown links inside the button group container:

```
<ul class="dropdown-menu">
  <li>...</li>
</ul>
```

5. Verify that you have linked to both jQuery and the Bootstrap JavaScript file at the bottom of the document, and then test it in your web browser.

As shown in Listing 12.4, the HTML looks more complicated than it really is. Figure 12.5 shows what it looks like.

LISTING 12.4   A Split Button Dropdown Menu

```
<div class="btn-group">
  <button class="btn btn-primary">Choose One...</button>
  <button class="btn btn-primary dropdown-toggle"
          data-toggle="dropdown">
    <span class="caret"></span>
    <span class="sr-only">Trigger Dropdown Menu</span>
  </button>
    <ul class="dropdown-menu">
      <li class="dropdown-header">Section 1</li>
      <li><a href="#">Item 1</a></li>
      <li><a href="#">Item 2</a></li>
      <li class="disabled"><a href="#">Item 3</a></li>
      <li class="divider"></li>
      <li class="dropdown-header">Section 2</li>
      <li><a href="#">Item 4</a></li>
      <li><a href="#">Item 5</a></li>
    </ul>
</div>
```

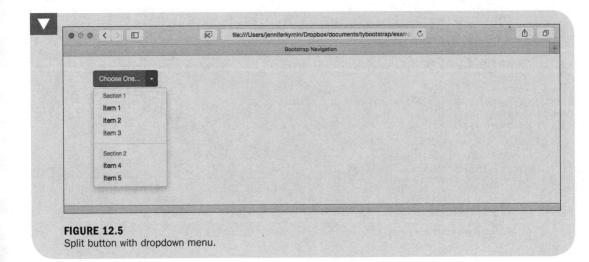

FIGURE 12.5
Split button with dropdown menu.

Dropup Variation

Sometimes you don't want the menu to drop below the button or toggle element, but rather to go up above it. Bootstrap provides a way to do this too with the class .dropup.

All you do is apply the .dropup class to the button group container in the same way you would apply the .dropdown class. Listing 12.5 shows a simple dropup menu.

LISTING 12.5   A Simple Dropup Menu

```
<div class="btn-group dropup">
  <button class="btn btn-default dropdown-toggle"
          data-toggle="dropdown">
    Choose One...
    <span class="caret"></span>
  </button>
  <ul class="dropdown-menu">
    <li class="dropdown-header">Section 1</li>
    <li><a href="#">Item 1</a></li>
    <li><a href="#">Item 2</a></li>
    <li class="disabled"><a href="#">Item 3</a></li>
    <li class="divider"></li>
    <li class="dropdown-header">Section 2</li>
    <li><a href="#">Item 4</a></li>
    <li><a href="#">Item 5</a></li>
  </ul>
</div>
```

CAUTION

Leave Space for Dropup Menus

You should not use dropup menus at the top of web pages because the top of the window could cut off the menu. Browsers do not add scrollbars when the menus appear, so they are impossible to see or click on if they open beyond the window frame.

Be sure to add text or other content above the HTML in Listing 12.5 so that it's not cut off by the browser window as the caution notes.

One thing to note about using the `.dropup` class is that it works best on `.btn-group` elements. If you want your dropup menu to appear without being in a button group, you simply add the `.btn-group` class to the container but do not add multiple buttons. Listing 12.5 shows this.

Navbars

Navigation in web pages is often found in the form of navigation bars, or *navbars*, and Bootstrap adds some specific features for navbars. These work as headers for your website or web application. Bootstrap navbars

- ▶ Start out collapsed in mobile views and can be toggled on and off

- ▶ Include a place for your brand text or image

- ▶ Can include buttons, button groups, dropdowns, and non-navigation links

- ▶ Allow for forms including search boxes

- ▶ Can be aligned to the right or left and fixed to the top or the bottom of the screen as well as positioned statically on the page

- ▶ Have an inverted option for different colors

When you are building navbars in Bootstrap, it's important to remember that Bootstrap is a mobile first framework. So everything you build is first designed for a mobile device, and then features for larger devices are added. Figure 12.6 shows the navigation in Listing 12.6 in a narrow window and a full-sized window.

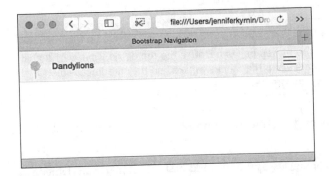

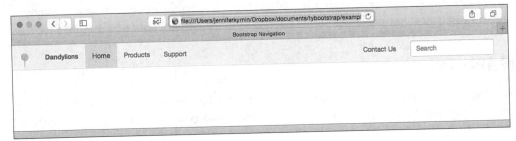

FIGURE 12.6
The same navigation in a narrow window and a wide one.

LISTING 12.6   The HTML for a Navbar

```
<nav class="navbar navbar-default">
  <div class="container-fluid">
    <div class="navbar-header">
      <button type="button" class="navbar-toggle collapsed"
              data-toggle="collapse" data-target="#collapsedNav">
      <span class="sr-only">Toggle navigation</span>
      <span class="icon-bar"></span>
      <span class="icon-bar"></span>
      <span class="icon-bar"></span>
      </button>
      <a href="#" class="navbar-brand">
      <img src="images/dandylion-logo.png"
           style="height:100%; width: auto;" alt="Dandylions"/>
      </a>
      <p class="navbar-text"><strong>Dandylions</strong></p>
    </div>

    <div class="collapse navbar-collapse" id="collapsedNav">
      <ul class="nav navbar-nav">
        <li class="active"><a href="#">Home</a></li>
        <li><a href="">Products</a></li>
```

```
    <li><a href="">Support</a></li>
  </ul>
  <form class="navbar-form navbar-right" role="search">
    <div class="form-group">
      <input type="text" class="form-control"
             placeholder="Search">
    </div>
  </form>
  <ul class="nav navbar-nav navbar-right">
    <li><a href="">Contact Us</a></li>
  </ul>
  </div>
  </div>
</nav>
```

CAUTION

Watch Your Navbar Length

Bootstrap cannot tell how wide your navigation is going to be, and if it gets wider than the current window can handle, it will wrap to a second line. To prevent this, you can remove items from your navigation, reduce the width of the items in the navigation, display certain items only at larger widths with responsive utility classes (see Hour 13, "Bootstrap Utilities"), or change when your navbar switches from collapsed to horizontal mode with CSS.

The collapsing navigation is part of the navbar collapse plugin and is included in the Bootstrap JavaScript file. Be sure to include the collapse plugin in the Bootstrap JavaScript on your page; otherwise, your navigation will not toggle in mobile browsers.

To create a navbar in Bootstrap, you first add a `<nav>` element with the class `.navbar`. The most common type of navbar uses the `.navbar-default` class. Inside the `<nav>` element, you place a container div with either the `.container` or `.container-fluid` class on it to define where the navigation should display. Finally, you place your branding elements, buttons, links, form fields, and other navigation tools in your navbar as you like.

Navbar Headers and Brands

A navbar header is a section of the navigation that will be seen by most users. It usually contains things like branding and logos, buttons to toggle the navigation on and off, and other critical items. You define your navbar header with the class `.navbar-header`.

You can add a logo or brand to your Bootstrap navbar with the `.navbar-brand` class. You can use either text or an image for your navbar brand. Listing 12.7 shows a basic navbar brand using text.

LISTING 12.7   Navbar Brand

```
<nav class="navbar navbar-default">
  <div class="container-fluid">
    <div class="navbar-header">
      <a href="#" class="navbar-brand">Dandylions</a>
    </div>
  </div>
</nav>
```

Toggling the Navigation

To be responsive, navigation is usually collapsed and hidden in mobile browsers, and that means you need a way to let mobile users see the navigation. It's best to group the toggle button with the brand so that mobile users get a better experience. You group them inside the `.navbar-header` you learned about previously.

You can create a toggle button however you want, but Bootstrap makes it easy to create a "hamburger" icon using only HTML and CSS. In Listing 12.8 you can see three span tags that read `<span class="icon-bar"></span>`. Each one of these creates a small line in the button, and three stacked creates the hamburger icon.

LISTING 12.8   Building a Navigation Toggle Button

```
<button type="button" class="navbar-toggle collapsed"
        data-toggle="collapse" data-target="#collapsedNav">
  <span class="sr-only">Toggle navigation</span>
  <span class="icon-bar"></span>
  <span class="icon-bar"></span>
  <span class="icon-bar"></span>
</button>
```

The button uses the `.navbar-toggle` and `.collapsed` classes to indicate that this is a toggle button and that it should be shown only when the navbar is in a collapsed state. Here are two other attributes that you should be aware of:

▶ `data-toggle="collapse"`—This is a data field that tells the JavaScript that this is a collapsed element. The script uses this field to know when to collapse and uncollapse the menu.

▶ `data-target="#collapsedNav"`—This is a data field that tells the JavaScript which elements to collapse and uncollapse. Later in the menu, there will be a container element with the `id` of `collapsedNav`. This is what will appear and disappear when the navigation is toggled on and off.

The last thing you need to create a toggleable navigation system is the navigation that will be turned on and off. Add a container element `<div>` with the classes `.collapse` and `.navbar-collapse`. Be sure to give it an `id` that matches the `data-target` you listed in your toggle button. Then place all the navigation elements you want toggleable in that container. Figure 12.7 shows navigation that has been opened with the toggle.

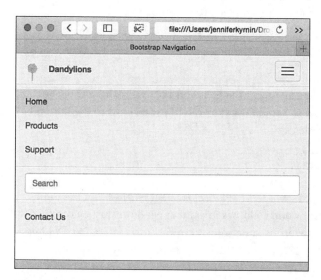

FIGURE 12.7
An open navigation bar.

Links, Text, Buttons, and Forms in Navbars

Most navbars are made up of links, buttons, and dropdown menus, and you can include all those things in your Bootstrap navbars as well as plain text.

Add links to your navbar with an unordered list. Give the list the classes `.nav` and `.navbar-nav` so they display correctly in the navbar. Listing 12.9 shows the HTML for a collapsible navbar section with three links.

LISTING 12.9   Collapsible Navbar with Three Links

```
<div class="collapse navbar-collapse" id="collapsedNav">
  <ul class="nav navbar-nav">
    <li class="active"><a href="#">Link</a></li>
    <li><a href="#">Link</a></li>
    <li><a href="#">Link</a></li>
  </ul>
</div>
```

The `collapse` function won't do anything other than disappear in larger screens until you have the JavaScript included in a complete Bootstrap page. See Listing 12.11 for a full listing.

Although you can simply add text to your navbar, this can result in some odd-looking layouts. So, Bootstrap provides the style `.navbar-text` to give your text the correct color and leading. In addition, a `.navbar-link` class is available for links in your navbar that aren't part of the navigation. This gives them a more standard link look and feel. For example:

```
<p class="navbar-text">
  We love <a href="" class="navbar-link">Weeds</a>!
</p>
```

You add buttons the same way you add buttons anywhere else on your web page—with the `.btn` and `.btn-style` classes. But you also should use the `.navbar-btn` class so the button is vertically centered in the navbar. For example:

```
<button class="btn btn-default navbar-btn"
        type="button">Click Me</button>
```

Forms are added in a similar way, using the `.navbar-form` class. This ensures that the form is aligned correctly vertically in the navbar and collapses in smaller windows. Just be careful that you check your form fields in several browsers because some form controls require fixed widths to display correctly in the navbar. Listing 12.10 shows the HTML for a form and button.

LISTING 12.10   A Form and Button in a Navbar

```
<nav class="navbar navbar-default">
  <div class="container-fluid">
    <div class="collapse navbar-collapse" id="collapsedNav">
      <form class="navbar-form navbar-right" role="search">
        <div class="form-group">
          <input type="text" class="form-control"
                 placeholder="Search">
        </div>
      </form>
      <button type="button"
              class="btn btn-default navbar-btn navbar-right">
        Contact Us
      </button>
    </div>
  </div>
</nav>
```

Figure 12.8 shows a navbar with several different components, and Listing 12.11 shows the HTML for it.

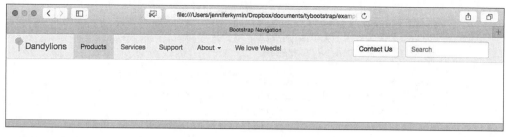

FIGURE 12.8
A navbar with multiple components.

LISTING 12.11   HTML for a Navbar with Multiple Components

```
<!DOCTYPE html>
<html lang="en">
  <head>
    <meta charset="utf-8">
    <meta http-equiv="X-UA-Compatible" content="IE=edge">
    <meta name="viewport"
          content="width=device-width, initial-scale=1">
    <title>Bootstrap Navigation</title>

    <!-- Bootstrap -->
    <link href="css/bootstrap.min.css" rel="stylesheet">
    <!-- HTML5 shim and Respond.js for IE8 support of HTML5
    elements and media queries -->
    <!-- WARNING: Respond.js doesn't work if you view the page
    via file:// -->
    <!--[if lt IE 9]>
      <script
src="https://oss.maxcdn.com/html5shiv/3.7.2/html5shiv.min.js">
      </script>
      <script
src="https://oss.maxcdn.com/respond/1.4.2/respond.min.js"></script>
    <![endif]-->
    <style>
    img#dandylionLogo {
      height:100%; width: auto; display: inline; margin-top: -10px;
    }
    </style>
  </head>
  <body>
    <nav class="navbar navbar-default">
      <div class="container-fluid">
        <div class="navbar-header">
          <button type="button" class="navbar-toggle collapsed"
                  data-toggle="collapse"
                  data-target="#collapsedNav">
```

```
        <span class="sr-only">Toggle navigation</span>
        <span class="icon-bar"></span>
        <span class="icon-bar"></span>
        <span class="icon-bar"></span>
      </button>
      <a href="#" class="navbar-brand">
        <img src="images/dandylion-logo.png" alt="Dandylion"
             id="dandylionLogo" />Dandylions
      </a>
    </div>

    <div class="collapse navbar-collapse" id="collapsedNav">
      <ul class="nav navbar-nav">
        <li class="active"><a href="#">Products</a></li>
        <li><a href="#">Services</a></li>
        <li><a href="#">Support</a></li>
      <li class="dropdown">
      <a href="#" class="dropdown-toggle"
         data-toggle="dropdown" role="button"
         aria-expanded="false">
        About <span class="caret"></span>
      </a>
      <ul class="dropdown-menu" role="menu">
        <li><a href="#">Articles</a></li>
        <li><a href="#">Related Sites</a></li>
      </ul>
      </li>
      </ul>
      <p class="navbar-text">
        We love <a href="" class="navbar-link">Weeds</a>!
      </p>
      <form class="navbar-form navbar-right" role="search">
        <div class="form-group">
          <input type="text" class="form-control"
                 placeholder="Search">
        </div>
      </form>
      <button type="button"
              class="btn btn-default navbar-btn navbar-right">
        Contact Us
      </button>
      </div>
    </div>
  </nav>

<script src="http://code.jquery.com/jquery-latest.js"></script>
<script src="js/bootstrap.min.js"></script>

  </body>
</html>
```

Changing the Navbar Colors and Alignment

Bootstrap navbars default to a light gray background color with darker gray highlights. But an alternative dark version exists that you can use by switching to the inverted navbar with the `.navbar-inverse` class. Figure 12.9 shows what that looks like.

FIGURE 12.9
A navbar with inverted colors.

Put the `.navbar-inverse` class on the container element for the navbar. This will change the colors to a black navigation bar with dark gray highlights. You will learn how to modify Bootstrap to use other colors in Hour 21, "Customizing Bootstrap and Your Bootstrap Website."

You also can adjust where the navbar displays and whether it scrolls with the content. Having navigation fixed to the top or bottom of the screen can be useful for sites that have a lot of information or extremely long pages. Bootstrap lets you easily adjust your navigation position with these classes:

- ▶ `.navbar-static-top`—Removes any padding and margin around the navbar to position it directly at the top of the page. The navbar will scroll with the rest of the page.

- ▶ `.navbar-fixed-top`—Removes the padding and margin around the top of the navbar and positions it at the top of the window. Content will then scroll beneath it, but it will never leave the window.

- ▶ `.navbar-fixed-bottom`—Removes the padding and margin around the bottom of the navbar and positions it at the bottom of the window. Content then scrolls beneath it, but it will never leave the window.

When you fix the navbar to the top of the window, you must add some CSS to your body tag so it won't be covered by the navigation. The navbar starts out at 50px tall, but you should test values that work for your design:

```
body { padding-top: 70px; }
```

As shown in Figure 12.10, scrolling pushes the content underneath the fixed navbar.

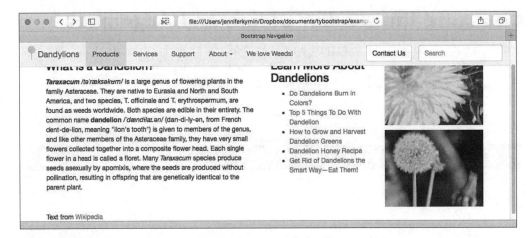

FIGURE 12.10
Scrolling content with a fixed navbar.

You also can position the elements inside your navbar with the `.navbar-left` and `.navbar-right` classes. This adds a CSS `float` to the property in the indicated direction. Be aware that there is a limitation on the way `.navbar-right` elements are positioned, and the margins might look off. Be sure to test this in your browsers. This component is being looked at for revision in Bootstrap version 4.

Breadcrumbs and Pagination

Two other common forms of navigation are breadcrumbs and pagination. These are typically used as extra navigation for large sites or large documents.

Breadcrumbs create a site hierarchy that allows customers to discover where they are quickly and easily. Bootstrap automatically adds separators to your breadcrumb links. Turn an ordered list into breadcrumbs with the class `.breadcrumb`. You can then indicate the current page with the `.active` class. Listing 12.12 shows how.

LISTING 12.12   Bootstrap Breadcrumbs

```
<ol class="breadcrumb">
  <li><a href="">Dandelions</a></li>
  <li><a href="">Recipes</a></li>
  <li><a href="">Drinks</a></li>
  <li class="active">Dandelion Wine</li>
</ol>
```

Pagination gives links to multiple pages of content within your site or application. There are two ways to do it: standard pagination and the pager alternative.

To create pagination, write a list of links to the pages and add the class .pagination to it. Most pagination includes icons at the beginning and end of the list to enable customers to simply move to the next or previous page. Add those by adding previous and next links with the icons or characters you want to use. Listing 12.13 shows a standard Bootstrap pagination list.

LISTING 12.13   Standard Bootstrap Pagination

```
<ul class="pagination">
  <li>
    <a href="" aria-label="Previous">
      <span class="glyphicon glyphicon-arrow-left"></span>
    </a>
  </li>
  <li><a href="">1</a></li>
  <li><a href="">2</a></li>
  <li><a href="">3</a></li>
  <li><a href="">4</a></li>
  <li><a href="">5</a></li>
  <li>
    <a href="" aria-label="Next">
      <span class="glyphicon glyphicon-arrow-right"></span>
    </a>
  </li>
</ul>
```

You can make your pagination links larger with the .pagination-lg class and smaller with the .pagination-sm class. By using the .active and .disabled classes, you can highlight the current page and deemphasize pages that don't work (like the "previous" page on the first page of a document).

If you don't want to list every page or you just need a simpler "next/previous" style pagination, Bootstrap provides the .pager class. Put this class on the unordered list, and then the two list items should be "Previous" and "Next." If you add the classes .previous and .next to the list items, this will align them to the left and right sides of the container. Listing 12.14 shows a pager list with the links aligned to the sides.

LISTING 12.14   Pager List with Links to the Side

```
<ul class="pager">
  <li class="previous"><a href="">
    <span class="glyphicon glyphicon-arrow-left"></span> Previous
  </a></li>
  <li class="next"><a href="">
    Next <span class="glyphicon glyphicon-arrow-right"></span>
  </a></li>
</ul>
```

Figure 12.11 shows some breadcrumbs and pagination in Bootstrap.

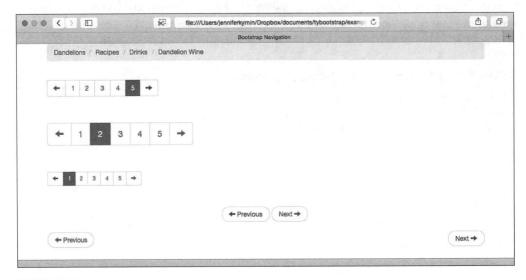

FIGURE 12.11
Breadcrumbs and pagination in Bootstrap.

List Groups

List groups are a component you can use to display simple and complex lists with your own custom content. They also can be used to create interesting vertical navigation.

To create a list group, build a standard HTML list and add the .list-group class to the container—either <ul> or <ol>. Then each <li> item in the list gets the .list-group-item class. Figure 12.12 shows what a simple list group looks like.

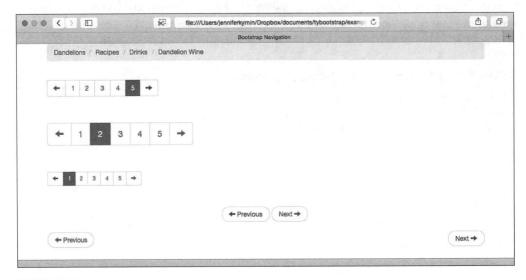

FIGURE 12.12
A simple list group.

If you add badges to your list group (learn about badges in Hour 6, "Labels, Badges, Panels, Wells, and the Jumbotron"), they will be automatically aligned to the right. Listing 12.15 shows what the HTML for a list group with badges looks like.

LISTING 12.15   A List Group with Badges

```
<ul class="list-group">
  <li class="list-group-item"><em>Taraxacum albidum</em>
    <span class="badge">14</span></li>
  <li class="list-group-item"><em>Taraxacum aphrogenes</em>
    <span class="badge">3</span></li>
  <li class="list-group-item"><em>Taraxacum brevicorniculatum</em>
    <span class="badge">5</span></li>
  <li class="list-group-item"><em>Taraxacum californicum</em>
    <span class="badge">27</span></li>
  <li class="list-group-item"><em>Taraxacum centrasiaticum</em>
    <span class="badge">0</span></li>
</ul>
```

You can create a navigation menu by turning a group of links into a list group. You don't even need the `<ul>` or `<li>` tags (although it's more accessible to use them). Listing 12.16 shows how to turn a group of links into a navigation list group.

LISTING 12.16   A Group of Links as a List Group

```
<div class="list-group">
  <a href="" class="list-group-item"><em>Taraxacum albidum</em></a>
  <a href="" class="list-group-item active"><em>Taraxacum
  aphrogenes</em></a>
  <a href="" class="list-group-item"><em>Taraxacum
  brevicorniculatum</em></a>
  <a href="" class="list-group-item"><em>Taraxacum
  californicum</em></a>
  <a href="" class="list-group-item"><em>Taraxacum
  centrasiaticum</em></a>
</div>
```

You can adjust how the list group items look with the `.disabled` and `.active` classes. Several contextual classes exist that you add to each list group item:

- ▶ `.list-group-item-info`—Indicates the content is informative

- ▶ `.list-group-item-success`—Indicates a successful or positive item

- ▶ `.list-group-item-danger`—Indicates a dangerous or negative item

- ▶ `.list-group-item-warning`—Indicates something that might be dangerous or difficult

As with all contextual classes, you should always provide alternatives for nonvisual browsers and screen readers if the classes provide information critical to the content.

Where list groups shine as vertical navigation is when you add the `.list-group-item-heading` and `.list-group-item-text` classes. These enable you to create linked blocks with headlines and descriptive text, as in Figure 12.13.

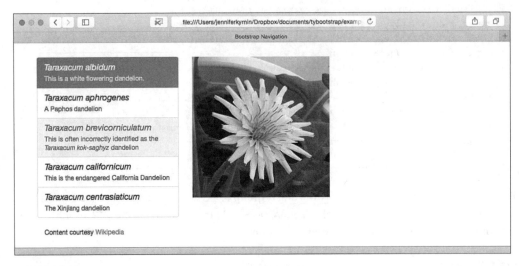

FIGURE 12.13
A fancy list group.

As shown in Figure 12.13, the list group describes several dandelion species, with the active item shown as a picture and the commonly misidentified one highlighted with a warning color. This gives you a great navigation structure for a photo gallery where you want to provide additional information about each navigation item. The HTML for Figure 12.13 is in Listing 12.17.

LISTING 12.17   A Fancy List Group

```
<div class="container">
  <div class="row">
  <div class="list-group col-md-4">
    <a href="" class="list-group-item active">
      <h4 class="list-group-item-heading"><em>Taraxacum
      albidum</em></h4>
      <p class="list-group-item-text">This is a white flowering
      dandelion.</p>
    </a>
    <a href="" class="list-group-item">
      <h4 class="list-group-item-heading"><em>Taraxacum
      aphrogenes</em></h4>
```

```
      <p class="list-group-item-text">A Paphos dandelion</p>
   </a>
   <a href="" class="list-group-item list-group-item-warning">
      <h4 class="list-group-item-heading"><em>Taraxacum
      brevicorniculatum</em></h4>
      <p class="list-group-item-text">This is often incorrectly
      identified as the <em>Taraxacum kok-saghyz</em> dandelion</p>
   </a>
   <a href="" class="list-group-item">
      <h4 class="list-group-item-heading"><em>Taraxacum
      californicum</em></h4>
      <p class="list-group-item-text">This is the endangered
      California Dandelion</p>
   </a>
   <a href="" class="list-group-item">
      <h4 class="list-group-item-heading"><em>Taraxacum
      centrasiaticum</em></h4>
      <p class="list-group-item-text">The Xinjiang dandelion</p>
   </a>
   </div>
   <img src="images/T_albidum01.jpg" alt="T. albidum"
      class="col-md-4">
   <p class="col-md-12">Content courtesy
   <a href="http://en.wikipedia.org/wiki/Taraxacum">Wikipedia</a>
   </div>
</div>
```

Summary

This hour covered the various ways you can create navigation and other lists of links in Bootstrap. You learned about the standard navigation elements such as buttons and links as well as dropdown menus. You also learned how to create navigation bars or navbars.

Navbars add a lot of functionality to your web pages with things like headers and brands as well as features in the navbars such as links, buttons, forms, and text. You also learned how to change the alignment of the navbars as well as the items in them and how to change the color from gray to black. In addition, this hour covered less standard navigation structures like pagination, breadcrumbs, and list groups. All the CSS classes from Hour 12 are listed in Table 12.1.

TABLE 12.1 Bootstrap Classes for Navigation Elements

CSS Class	Description
.breadcrumb	Indicates the list is a breadcrumb list.
.caret	Displays a caret icon.
.collapse	Indicates that the navbar element should be part of the collapsible group.
.collapsed	The element should be displayed collapsed except for mobile browsers.
.divider	Places a divider in the dropdown menu.
.dropdown	Indicates the element contains a dropdown menu list.
.dropdown-header	A header for a dropdown menu list.
.dropdown-menu	The dropdown menu list.
.dropdown-menu-left	Positions the dropdown menu on the left side of the container.
.dropdown-menu-right	Positions the dropdown menu on the right side of the container.
.dropdown-toggle	Indicates the element should turn a dropdown menu on and off.
.dropup	The element contains a menu that should open above the item rather than drop down below.
.icon-bar	Displays a bar or dash line icon.
.list-group	The element contains a list group.
.list-group-item	The element is a list group item.
.list-group-item-danger	The list group item is dangerous or negative in some way.
.list-group-item-heading	A headline or heading for a list group item.
.list-group-item-info	The list group item provides information.
.list-group-item-success	The list group item is successful or positive in some way.
.list-group-item-text	A block of text in a list group item.
.list-group-item-warning	The list group item is possibly negative or dangerous.
.nav	A navigation element.
.nav-justified	A navigation item that is sized to be the same size as all the navigation items in the list.
.nav-pills	A navigation list that is displayed as buttons or pills.
.nav-tabs	A navigation list that is displayed as tabs.
.navbar	A navigation bar container.
.navbar-collapse	The section of the navigation bar that will collapse on mobile or small screen devices.

CSS Class	Description
.navbar-default	The standard look and feel navbar.
.navbar-fixed-bottom	Position the navbar at the bottom of the window with content scrolling below it.
.navbar-fixed-top	Position the navbar at the top of the window with content scrolling below it.
.navbar-form	A form element inside a navbar.
.navbar-header	A header section inside a navbar.
.navbar-inverse	Switch the standard colors from gray to black.
.navbar-link	Creates a standard link inside a navbar.
.navbar-nav	Indicates the actual navigation inside the navbar.
.navbar-static-top	Removes margins and padding on navbars to position them right at the top of the window.
.navbar-text	Styles standard text inside a navbar.
.navbar-toggle	Indicates the button or link that toggles the navbar on and off in small screen devices.
.next	Puts the button on the far right in a .pager list.
.pager	Creates a small next/previous pagination button scheme.
.pagination	Indicates the list is a pagination scheme.
.pagination-lg	Creates larger buttons in a pagination list.
.pagination-sm	Creates smaller buttons in a pagination list.
.previous	Puts the button on the far left in a .pager list.

Workshop

The workshop contains quiz questions to help you process what you've learned in this hour. Try to answer all the questions before you read the answers.

Q&A

Q. What is the difference between a `nav` element and a `navbar` element?

A. These two elements are very similar. The easiest way to think of them is this: the `nav` class is for any navigation structure you might have on a page, from a table of contents for the page itself to the full site navigation. The `navbar` class is used almost exclusively for the primary site navigation. Although it can be used in other locations, it is designed to be positioned at the top or bottom of the page.

Q. I use Bootstrap 2 and noticed that the `.nav-list` and `.nav-header` classes have been removed. What do I use instead?

A. There isn't a direct equivalent in Bootstrap 3, but using list groups as detailed previously or panel groups as detailed in Hour 6 are your best choices.

Quiz

1. What elements can you put `.nav` classes on?

   a. Only the `<nav>` element

   b. Only the `<div>` element

   c. Only `<nav>` and `<div>` elements

   d. Any block-level element

2. What is the difference between a `.nav-pill` and a `.nav-tab`?

   a. The pill class creates a button for each navigation element, and the tab creates a tab.

   b. The pill creates a button, and the tab creates a text link.

   c. The pill creates a text link, and the tab creates a button.

   d. The pill creates a menu, and the tab creates a tab.

3. What does the `.dropdown-header` class do?

   a. Creates a header describing the dropdown menu.

   b. Creates a header element in the navigation that indicates a dropdown menu.

   c. Creates a header inside the dropdown menu.

   d. Nothing; it's not a valid Bootstrap class.

4. True or False: This is the correct HTML to create a drop-up menu: `<button class="btn dropdown dropup" type="button">...</button>`

5. How is a split dropdown different from a button group?

   a. Button groups cannot contain dropdown menus.

   b. Button groups must contain more than two buttons.

   c. Split dropdowns do not use buttons.

   d. They are not different.

6. Which of the following elements is allowed in a navbar?

   a. Buttons

   b. Forms

   c. Text

   d. All of the above

   e. None of the above

7. Why should you create collapsible navigation?

   a. Navigation that collapses in smaller devices is more responsive and easier to use.

   b. Navigation that collapses takes up less space, so there's more room for content.

   c. Collapsed navigation is what most people expect from web designs.

   d. You should not create collapsible navigation as it is not mobile first.

8. Which HTML tags should you put the `.breadcrumb` class on?

   a. `<div>`

   b. The `<ul>` or `<ol>` tags

   c. `<p>`

   d. Any HTML tag you want to

9. Which of the following will create a pagination system for your page?

   a. Method 1:

```
<ul class="pagination">
  <li><a href="" aria-label="Previous">&lt;</a></li>
  <li><a href="">1</a></li>
  <li><a href="">2</a></li>
  <li><a href="" aria-label="Next">&gt;</a></li>
</ul>
```

   b. Method 2:

```
<ul class="paginate">
  <li>
    <a href="" aria-label="Previous">
      <span class="glyphicon glyphicon-arrow-left"></span>
    </a>
  </li>
  <li>
    <a href="" aria-label="Next">
      <span class="glyphicon glyphicon-arrow-right"></span>
    </a>
  </li>
</ul>
```

   c. Method 3:

```
<ul class="pager">
  <li class="previous"><a href="">Previous</a></li>
  <li class="next"><a href="">Next</a></li>
</ul>
```

    d. A and B

    e. A and C

    f. All of the above

10. Where do badges get positioned in list groups?

    a. They are positioned where you place them in the HTML.

    b. They are automatically positioned to the left.

    c. They are automatically positioned to the right.

    d. They are automatically hidden.

Quiz Answers

1. d. You can put the `.nav` class on any block-level element, but for accessibility and best practices you should use the `<nav>` element.

2. a. The pill creates a button, and the tab creates a tab.

3. c. Creates a header inside the dropdown menu.

4. False. You use the `.dropup` class instead of the `.dropdown` class, not alongside it.

5. d. They are not different. A split dropdown is simply a two-button button group where one button is a dropdown menu with only the caret showing.

6. d. All of the above.

7. a. Collapsible navigation is easier to use on mobile and small screen devices.

8. b. Although you can put the class on any HTML tag, best practices say you should use a list tag—either `<ul>` or `<ol>`.

9. e. Methods 1 and 3. If you look carefully, Method 2 has the class `.paginate`, which is not a valid Bootstrap class.

10. c. Badges are automatically positioned on the right side of the list item in a list group.

Exercises

1. Add a navbar with a dropdown menu as well as two other navigation elements. Test using a fixed navbar by positioning it either on the top or the bottom of your web page.

2. Create a navbar that is fixed to the bottom of the page. Be sure to test your navigation in as many devices as you can.

HOUR 13
Bootstrap Utilities

What You'll Learn in This Hour:

▶ How to change foreground and background colors
▶ How to align content on the page
▶ How to show and hide content in several ways
▶ How to embed media responsively
▶ Techniques for making your pages more accessible

Bootstrap offers a lot of utility classes to help you manage your pages and content more effectively. Some of the classes mentioned here have been covered in other areas of the book, but because they apply to more than just specific components, you will learn about them in more detail in this hour.

This hour will cover:

▶ How to change the foreground and background colors of elements on your pages

▶ How to float elements right and left as well as center them on the page

▶ How to display and hide content based on the device size or whether it's a printout, but also for design reasons

You also will learn how to embed media so that it flexes with the size of the device but stays in the correct aspect ratio. Plus, you'll learn some techniques for making your pages more accessible to people using assistive technology such as screen readers.

Helper Classes

Helper classes in Bootstrap extend the existing components, but more importantly, they give you additional functionality for standard HTML. In other words, you will be able to add colors, floats, and icons and show and hide content in a standard <p> tag—and not just in a Jumbotron (refer to Hour 6, "Labels, Badges, Panels, Wells, and the Jumbotron") or navbar (refer to Hour 12, "Creating Navigation Systems with Bootstrap").

Changing Colors

Several helper classes exist that you can use to change both the foreground (text) color and background color. These are called *contextual* classes because they can provide additional contextual clues about the content. In previous hours, you learned about contextual classes that are used with specific elements and components. These classes can be used with any text blocks to change the colors.

The text color classes change the foreground or text color and are as follows:

- `.text-muted`
- `.text-primary`
- `.text-success`
- `.text-info`
- `.text-warning`
- `.text-danger`

These classes can be applied directly to the container element holding the text you want to have that context. Or you can surround the text with a `<span>` tag to change the color of specific words. Listing 13.1 shows seven lines of text with the different classes and no class applied. Figure 13.1 shows how Listing 13.1 looks.

LISTING 13.1 Text Color Contextual Classes

```
<p>This paragraph is plain text, with no contextual class.</p>
<p class="text-muted">This paragraph is muted, class:
  <code>text-muted</code>.</p>
<p class="text-primary">This paragraph is primary, class:
  <code>text-primary</code>.</p>
<p class="text-success">This paragraph is success, class:
  <code>text-success</code>.</p>
<p class="text-info">This paragraph is information, class:
  <code>text-info</code>.</p>
<p class="text-warning">This paragraph is warning, class:
  <code>text-warning</code>.</p>
<p class="text-danger">This paragraph is danger, class:
  <code>text-danger</code>.</p>
```

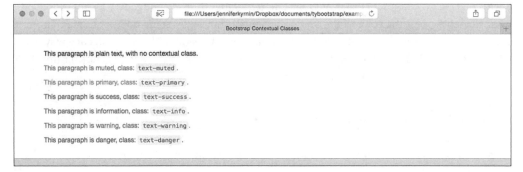

FIGURE 13.1
Text color contextual classes.

You can also use the text classes to style links. When customers hover over the link, the text darkens with the contextual color, just like non-styled links.

You also can style the background color on text with contextual classes. The background color classes are

▶ .bg-primary

▶ .bg-success

▶ .bg-info

▶ .bg-warning

▶ .bg-danger

Figure 13.2 shows what these classes look like.

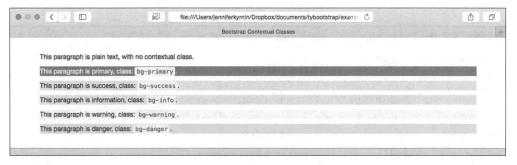

FIGURE 13.2
Background color contextual classes.

As with all other contextual classes, you need to remember that screen readers do not render visual elements. So, if the color of the text or background provides some additional meaning that is critical to the page, you should include it in some other way. The most effective way is to include a text block that is hidden from non-screen readers with the `.sr-only` class. Listing 13.2 gives a sample of how you can do it.

LISTING 13.2   Making Contextual Classes Accessible

```
<p class="bg-danger">
  <span class="sr-only">Danger!</span> Tigers will eat people if
  you annoy them.
</p>
```

CAUTION

CSS Specificity Can Get in the Way

Sometimes contextual classes can be overridden by other CSS classes that are more specific. The best solution is to surround the text with `<span>` tags holding the `text-*` classes and surround the text with `<div>` tags holding the `bg-*` classes.

Icons

Other than Glyphicons (refer to Hour 10, "Images, Media Objects, and Glyphicons"), there are a couple classes you can use to add other icons to the text.

The `.caret` class displays a caret that indicates dropdown functionality. It also reverses automatically when you use a dropup menu (refer to Hour 12).

Add this class by placing it on an empty `<span>` tag, as in Listing 13.3.

LISTING 13.3   Using the `.caret` Class

```
<span class="caret"></span>
```

The `.close` class is added to buttons to create a close button for modal dialog boxes and alerts. It adds a small gray *x* to the right side of the container. The *x* darkens when you mouse over it. Listing 13.4 shows how to add a `.close` icon button.

LISTING 13.4   Using the `.close` Class

```
<button type="button" class="close" aria-label="Close">
  <span aria-hidden="true">&times;</span>
</button>
```

Figure 13.3 shows the .caret and .close classes in action.

FIGURE 13.3
Carets and close icons.

Layout Classes

There are lots of ways you can adjust the layout of web pages with Bootstrap. These utility classes help you float your content on the page and center certain elements or even the entire page.

You can position content on the left and right with the .pull-left and .pull-right classes. You should not use these classes in navbars. Use .navbar-left and .navbar-right for those elements. Remember that if an element does not have some sort of width on it (either an explicit within CSS or a column width from the Bootstrap grid), floating it will do nothing because it already takes up the full width of the container.

TRY IT YOURSELF ▼

Create a Pull Quote from a Paragraph

Bootstrap lets you easily convert a paragraph into a fancy-looking pull-quote with just a few styles:

1. Open the page where you want the quotation in a web page editor.

2. Add a container <div>:

   ```
   <div class="container">
   </div>
   ```

3. Add one to three paragraphs of text inside the container.

4. Include the quotation as one paragraph. Place that quotation where you want it to be positioned on the page.

5. Add the class .col-sm-3 to the paragraph with the quote. This will resize it to be three columns wide.

6. Add the class `.pull-right` to that paragraph as well.

7. Create styles that make the quote larger; then add background and foreground colors and other styles to decorate the quote.

Listing 13.5 shows the HTML I used to create the page shown in Figure 13.4.

LISTING 13.5   Create a Pull Quote from a Paragraph

```
<!DOCTYPE html>
<html lang="en">
  <head>
    <meta charset="utf-8">
    <meta http-equiv="X-UA-Compatible" content="IE=edge">
    <meta name="viewport"
          content="width=device-width, initial-scale=1">
    <title>Pull Quote Paragraph</title>

    <!-- Bootstrap -->
    <link href="css/bootstrap.min.css" rel="stylesheet">
    <!-- HTML5 shim and Respond.js for IE8 support of HTML5
    elements and media queries -->
    <!-- WARNING: Respond.js doesn't work if you view the page
    via file:// -->
    <!--[if lt IE 9]>
      <script
src="https://oss.maxcdn.com/html5shiv/3.7.2/html5shiv.min.js">
      </script>
      <script
src="https://oss.maxcdn.com/respond/1.4.2/respond.min.js"></script>
    <![endif]-->
    <style>
    p.pull-right {
      border: solid green 3px;
      color: #F0E433;
      background-color: #025301;
      padding: 1em;
      margin-left: 1em;
      font-size: 1.5em;
    }
    </style>
  </head>
  <body>
    <p> </p>
    <div class="container">
      <p class="col-sm-3 pull-right">&#8220;If dandelions were hard
```

```
          to grow, they would be most welcome on any lawn.&#8221;<br>
          <span class="small">~Andrew V. Mason</span></p>
          <p>Taraxacum /təˈræksəkʉm/ is a large genus of flowering
          plants in the family Asteraceae. They are native to Eurasia
          and North and South America, and two species, T. officinale
          and T. erythrospermum, are found as weeds worldwide. Both
          species are edible in their entirety. The common name
          dandelion (/ˈdændɪlaɪ.ən/ dan-di-ly-ən, from French
          dent-de-lion, meaning "lion's tooth") is given to members of
          the genus and, like other members of the Asteraceae family,
          they have very small flowers collected together into a
          composite flower head. Each single flower in a head is called
          a floret. Many Taraxacum species produce seeds asexually by
          apomixis, where the seeds are produced without pollination,
          resulting in offspring that are genetically identical to the
          parent plant.</p>
          <p>Text courtesy
          <a href="http://en.wikipedia.org/wiki/Taraxacum">
            Wikipedia</a>.
          </p>
        </div>
    </body>
</html>
```

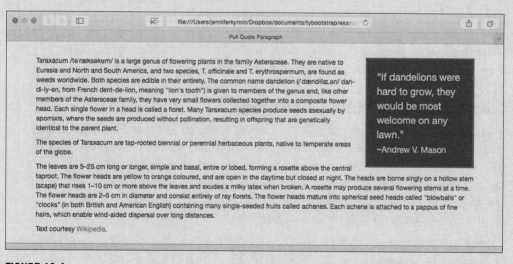

FIGURE 13.4
A page with a pull quote.

When you float elements, you need to be able to clear them so that content stops floating. Bootstrap provides the .clearfix class for that purpose. You apply the .clearfix class when you want the float to be cleared *after* that element. Be aware that this is slightly different from other clear fix styles you might find on the Web.

Bootstrap includes a class .center-block that centers any element as a block. It converts the element to display as a block-level element and then centers it with automatic margins on the left and the right.

CAUTION
Centered Blocks Must Have a Width

For blocks centered with the .center-block class to actually be centered, they need to have an explicit width. This means you need to include a style on the element that sets the width. You cannot use Bootstrap grid classes to set the width and center the element.

Listing 13.6 shows how you might center a paragraph with the .center-block class.

LISTING 13.6   Center a Paragraph

```
<p class="center-block" style="width:300px;">
  &#8220;If dandelions were hard to grow, they would be most
  welcome on any lawn.&#8221;<br>
  <span class="small">~Andrew V. Mason</span>
</p>
```

Notice that I set the width of the paragraph with a style attribute. I did this to make it easy to see, but it is a better practice to set the style as a class in your personal style sheet.

Displaying and Hiding Content

Several helper classes are available that you can use to display and hide content in different situations. Two that help accessibility are the .sr-only and .sr-only-focusable classes. The .sr-only class I've mentioned previously as a class to show the enclosed content only to screen readers. This hides it from standard browsers, but screen readers will still read it. Add the .sr-only-focusable to the element if you need it to show the element again when it's focused on. This is particularly useful for keyboard-only users.

Listing 13.7 shows how you might write a skip link for screen readers to skip navigation lists and go directly to the content that wouldn't be shown to a standard browser user.

LISTING 13.7 HTML for a Skip Link

```
<a class="sr-only sr-only-focusable" href="#content">Skip to main
content</a>
<!-- ... navigation to be skipped ... -->
<a name="content"></a>
```

You can use the .text-hide class to do image replacements. This creates accessible HTML content that is readable by screen readers but uses a graphic that is attractive to visual browsers. This is covered in more detail in Hour 7, "Bootstrap Typography."

A few classes allow you to show and hide block-level content directly on your page:

▶ .show—The content will be visible on the page to all devices, including screen readers.

▶ .hidden—The content will be removed from the page and not seen by any devices, including screen readers.

▶ .invisible—The content will be invisible on the page but still take up space in the flow of content.

Note that the .hide class has been deprecated as of Bootstrap 3.0.1, and you should not use it. Use .hidden, .invisible, or .sr-only instead.

Responsive Utilities

Bootstrap offers a number of utility classes you can use to show and hide content from specific devices, based on media queries.

As with the Bootstrap layout grids, you can target four standard sizes: extra-small (xs), small (sm), medium (md), and large (lg). You can define three types of element: block, inline, and inline-block. You can either hide the content (hidden) or show it (visible). These are combined to create the responsive classes.

The classes to hide elements are

▶ .hidden-xs—Hides the content from extra-small devices

▶ .hidden-sm—Hides the content from small devices

▶ .hidden-md—Hides the content from medium-sized devices

▶ .hidden-lg—Hides the content from large devices

The classes to show elements include how to display the content:

▶ `.visible-xs-block`—Display to extra-small devices as a block-level element

▶ `.visible-xs-inline`—Display to extra-small devices as an inline element

▶ `.visible-xs-inline-block`—Display to extra-small devices as an inline-block element

▶ `.visible-sm-block`—Display to small devices as a block-level element

▶ `.visible-sm-inline`—Display to small devices as an inline element

▶ `.visible-sm-inline-block`—Display to small devices as an inline-block element

▶ `.visible-md-block`—Display to medium-sized devices as a block-level element

▶ `.visible-md-inline`—Display to medium-sized devices as an inline element

▶ `.visible-md-inline-block`—Display to medium-sized devices as an inline-block element

▶ `.visible-lg-block`—Display to large devices as a block-level element

▶ `.visible-lg-inline`—Display to large devices as an inline element

▶ `.visible-lg-inline-block`—Display to large devices as an inline-block element

To use these classes, you can add them to the elements you want to show or hide. Or you can add a container around the content you want to modify the display.

CAUTION

Your Pages Should Use Progressive Enhancement

It can be tempting to create web pages that look completely different with different content for different devices. But this is not progressive enhancement. Progressive enhancement is when you create a site that is mobile first and then enhanced for the larger devices. This means that all the important content is displayed to everyone, and then features that improve the site are added to devices that can handle them. So you should be cautious about using the responsive classes unless you absolutely need them because they can make your site less progressively enhanced.

The classes `.visible-xs`, `.visible-sm`, `.visible-md`, and `.visible-lg` also exist, but they have been deprecated as of Bootstrap 3.2.0. Use the `.visible-*-block` classes instead. Table 13.1 shows how these classes display across different device sizes.

TABLE 13.1 How Responsive Classes Display on Different Device Sizes

Class	Extra-Small Devices	Small Devices	Medium Devices	Large Devices
`.hidden-xs`	Hidden	Visible	Visible	Visible
`.hidden-sm`	Visible	Hidden	Visible	Visible
`.hidden-md`	Visible	Visible	Hidden	Visible
`.hidden-lg`	Visible	Visible	Visible	Hidden
`.visible-xs-*`	Visible	Hidden	Hidden	Hidden
`.visible-sm-*`	Hidden	Visible	Hidden	Hidden
`.visible-md-*`	Hidden	Hidden	Visible	Hidden
`.visible-lg-*`	Hidden	Hidden	Hidden	Visible

Print Classes

Bootstrap also provides a few utility classes to help you show and hide content when your web pages are printed. This is useful for when you don't want to create a full print style sheet or you just need to show or hide a few things from printouts.

The print classes are

▶ `.hidden-print`—Hides the content from print view

▶ `.visible-print-block`—Displays the content as a block-level element on print views

▶ `.visible-print-inline`—Displays the content as an inline element on print views

▶ `.visible-print-inline-block`—Displays the content as an inline-block element on print views

To use these classes, simply add them to the elements you want to show or hide.

Here are a few things you should consider when deciding what to print:

▶ Images take up a lot of toner or ink, so you should hide images that aren't critical to the content, such as icons and navigation images.

▶ Links are not clickable on printouts, but you can add the URLs as `.visible-print-inline` elements as footnotes or even right next to the links.

▶ Navigation elements are not useful on printed web pages, so you should hide them.

▶ Advertising is also not useful on printouts.

▶ Including your site name, a small logo, and the page URL because a header and footer on printouts make the printouts more useful and act as an advertisement for your website.

It can be tempting to use the print classes to hide elements you don't want printed, like copyrighted images, from the printer. But remember that this is not foolproof. If you absolutely must not allow your content to be printed, you should consider not posting it to the Web.

Responsive Embed

Bootstrap has classes that help you embed videos and slideshows with dimensions based on the width of the container element. Add the .embed-responsive class to a container element around the media element (<iframe>, <embed>, <video>, or <object>). Then add the .embed-responsive-item class to the media element itself. Finally, define the video size with the .embed-responsive-16by9 or .embed-responsive-4by3 class on the container element. Listing 13.8 shows how this would look for a 16×9 video.

LISTING 13.8   A 16×9 Video in Bootstrap

```
<div class="embed-responsive embed-responsive-16by9">
  <video class="embed-responsive-item">
    <source src="...">
  </video>
</div>
```

Accessibility in Bootstrap

Bootstrap, as you have learned already, has a number of classes set up to help people using assistive technology such as screen readers. The following are a few things that Bootstrap recommends you include on all your pages to make them more accessible:

▶ **Skip links**— These are links placed at the very top of the web page that enable people using screen readers to skip to the main content. The Bootstrap classes .sr-only and .sr-only-focusable hide these links from standard browsers so they don't affect the design. To be accessible, your pages should include a skip link at the top of every page.

▶ **Contrasting colors**— Some of the default color combinations in Bootstrap do not contrast as well as they could. To be more accessible, you should adjust the colors to have higher contrast.

▶ **Nested headlines**— Although you can use any headlines (`<h1>` through `<h6>`) that you want, it's most accessible to use them like an outline. Use the `<h1>` tag for the primary headline for the document. Then use the subsequent tags in logical order.

There are a few other simple things you can do to keep your website accessible. The most important is providing alternatives to any technology or feature that might be difficult to use. Because Bootstrap is mobile first, a lot of this is taken care of for you. But you need to remember to provide alternative text for images, use multiple source files for videos, and give fallback options for any scripts on the site.

Summary

This hour covers a lot of diverse classes that provide extra functionality for Bootstrap. There are helper classes, responsive utility classes, classes for print, and classes for embedding videos responsively.

The helper classes are the most diverse. There are classes for changing the foreground and background colors of the page and classes to add simple icons to buttons and dropdown menus. There are also several classes for positioning your content on the page, as well as for displaying and hiding it.

The responsive utilities let you show and hide content based on device size. Print classes let you show and hide content when your web pages are printed, and responsive embed classes let you resize your videos keeping the correct aspect ratio.

This hour also explains how to keep your Bootstrap pages accessible. Table 13.2 lists the CSS classes explained in this hour.

TABLE 13.2 Utility Classes in Bootstrap

CSS Class	Description
.bg-danger	Changes the background color to indicate the element is "danger."
.bg-info	Changes the background color to indicate the element is "info."
.bg-primary	Changes the background color to indicate the element is "primary."
.bg-success	Changes the background color to indicate the element is "success."
.bg-warning	Changes the background color to indicate the element is "warning."

CSS Class	Description
.bg-danger	Changes the background color to indicate the element is "danger."
.caret	Adds a small icon indicating a dropdown menu. Automatically switches to an up-pointing icon for dropup menus.
.center-block	Centers block-level elements.
.close	Adds a small *x* icon indicating a close button.
.embed-responsive	Indicates an embedded element that resizes based on the container width.
.embed-responsive-16by9	Resizes the embedded media in the 16:9 ratio.
.embed-responsive-4by3	Resizes the embedded media in the 4:3 ratio.
.embed-responsive-item	Indicates the media item that should be embedded responsively.
.hidden-lg	Element is hidden in large devices.
.hidden-md	Element is hidden in medium devices.
.hidden-print	Element is hidden when the page is printed.
.hidden-sm	Element is hidden in small devices.
.hidden-xs	Element is hidden in extra small devices.
.pull-left	Element is floated to the left.
.pull-right	Element is floated to the right.
.sr-only	Displays content only to screen readers and assistive technology.
.sr-only-focusable	Displays the content again when it receives focus.
.text-danger	Changes the foreground (font) color to indicate "danger."
.text-hide	Hides the text to do image replacement.
.text-info	Changes the foreground (font) color to indicate "information."
.text-muted	Changes the foreground (font) color to indicate it's less important or muted.
.text-primary	Changes the foreground (font) color to indicate "primary."
.text-success	Changes the foreground (font) color to indicate "success."
.text-warning	Changes the foreground (font) color to indicate "warning."
.visible-lg-block	Element is visible as a block-level element on large devices.
.visible-lg-inline	Element is visible as an inline element on large devices.
.visible-lg-inline-block	Element is visible as an inline-block element on large devices.

CSS Class	Description
`.visible-md-block`	Element is visible as a block-level element on medium devices.
`.visible-md-inline`	Element is visible as an inline element on medium devices.
`.visible-md-inline-block`	Element is visible as an inline-block element on medium devices.
`.visible-print-block`	Element is visible as a block-level element in print.
`.visible-print-inline`	Element is visible as an inline element in print.
`.visible-print-inline-block`	Element is visible as an inline-block element in print.
`.visible-sm-block`	Element is visible as a block-level element on small devices.
`.visible-sm-inline`	Element is visible as an inline element on small devices.
`.visible-sm-inline-block`	Element is visible as an inline-block element on small devices.
`.visible-xs-block`	Element is visible as a block-level element on extra small devices.
`.visible-xs-inline`	Element is visible as an inline element on extra small devices.
`.visible-xs-inline-block`	Element is visible as an inline-block element on extra small devices.

Workshop

The workshop contains quiz questions to help you process what you've learned in this hour. Try to answer all the questions before you read the answers.

Q&A

Q. It seems like you say we shouldn't use the responsive utility classes, like `.hidden-sm` or `.visible-lg-block`. But why would Bootstrap have them if they are not a good idea to use?

A. Bootstrap strives to meet the needs of the majority of web designers. Although best practices indicate that pages should be designed mobile first, with progressive enhancement, and provide the same content to all visitors, not all designers want or need to do all those things.

My recommendation is to use them sparingly. But if you need to use them, don't feel like you can't. They are there to be used—just think about your customers before you make a website that is completely different for mobile users than for desktop users.

Q. Why are there so many different contextual classes for the different elements, such as `.btn-warning`, `.has-success`, and `.text-info`?

A. It seems like it should be possible to just have one contextual class that affects all the elements and components. But Bootstrap developers recognized that different elements have different style issues, so they developed specific contextual classes to meet those needs. When using contextual classes, it's best to use the class specific for the element or component you're using, and only use the general ones listed in this hour if there are no specific ones.

Q. The accessibility guidelines you bring up seem like a big hassle for a small number of people. Are they really that important?

A. Accessibility is going to be more important for some sites than others. Many governments have accessibility laws that websites must conform to, and if you're designing for a public sector site, you will need to follow accessibility guidelines. Of course, sites that cater to people with different capabilities are going to need to be accessible as well. But it's a good idea to make even mainstream, non-government websites accessible because you never know who's going to visit your site. There are only a few things you need to do to create a more accessible site, like skip links, alternate text for images, and using good contrasting colors for text. These make your site better for everyone who uses it—and that benefits everyone.

Quiz

1. A helper class that changes the text color is called what?

   a. A color class

   b. A context class

   c. A contextual class

   d. A help class

2. What happens when you use a class from question 1 to change the color of a link?

   a. The link changes color only on hover.

   b. The link changes color and gets a darker color on hover.

   c. The link changes color when it's active.

   d. Nothing; the link stays the same color as usual.

3. True or False: You do not need to do anything to make an element accessible, beyond adding the class, when a helper class adds critical information to the element.

   a. True. The class provides the information to all customers automatically.

   b. True. The class makes the element accessible.

   c. False. You need to add another class to the element to make it accessible.

   d. False. You need to add more information for assistive technology to make it accessible.

4. Which of the following is the correct way to add a close icon?

    a. `<button type="button" class="close btn">`

    b. `<button type="close button" class="btn">`

    c. `<span class="close"></span>`

    d. `<div class="close"></div>`

5. How do you float navigation bar elements to the right?

    a. `.nav-right`

    b. `.navbar-right`

    c. `.pull-right`

    d. `.right`

6. What does `<p class="pull-right">` do?

    a. Creates a right aligned paragraph, with text inside left aligned.

    b. Creates a right aligned paragraph, with text inside right aligned.

    c. Creates a left aligned paragraph, with text inside left aligned.

    d. Creates a left aligned paragraph, with text inside right aligned.

    e. Nothing. The paragraph will act like every other paragraph around it.

7. Which of the following is a valid way to remove content from the flow with Bootstrap?

    a. `.hide`

    b. `.hidden`

    c. `.invisible`

    d. All of the above

8. How could you hide an element from only very small mobile devices?

    a. `.hidden`

    b. `.hidden-xs`

    c. `.hidden-sm`

    d. `.hide-xs`

9. Why would you hide content from a printout?

    a. To reduce the amount of ink or toner used

    b. To remove elements that don't work in print

    c. To remove advertising or other elements that aren't needed on a printout

    d. Both a and b

    e. All of the above

10. Which of the following is not an accessibility issue?

    a. Blurry images that are hard to understand.

    b. Low contrast colors.

    c. Headlines are not in outline order.

    d. Video that does not have alternative text.

Quiz Answers

1. c. A contextual class because it provides contextual information.

2. b. The link changes color and gets a darker color on hover.

3. d. False. You need to add more information for assistive technology to make it accessible.

4. a. `<button type="button" class="close btn">`

5. b. `.navbar-right`

6. e. Nothing. Because the paragraph has no width set, it will take up the full width of the container, and so act like every other paragraph around it.

7. b. `.hidden` removes the content by setting the display to none. `.hide` has been deprecated. `.invisible` only sets the visibility to hidden.

8. b. `.hidden-xs`

9. e. All of the above.

10. a. Blurry images are bad on a web page, but they don't make the page any more or less accessible simply because they are blurry.

Exercises

1. Add the print classes to a web page to make your page more print-friendly.

2. Check your website with an accessibility checker, such as one found on the W3C: http://www.w3.org/WAI/ER/tools/. Look at some of the suggestions and see whether you can implement as many as possible to make your pages more accessible.

Using Bootstrap JavaScript Plugins

What You'll Learn in This Hour:

▶ How to add Bootstrap JavaScript plugins

▶ How to set options for the plugins

▶ How to use the JavaScript API

▶ How to prevent some common problems with the Bootstrap plugins

In Part II, "Building and Managing Web Pages with Bootstrap," you learned a lot about the various CSS styles and components that Bootstrap provides. But Bootstrap also includes more than 10 JavaScript plugins you can use to add interactivity and dynamic elements to your website. In this hour you learn how to use the plugins as well as the API so you can program your own plugins with special features. In the following hours in Part III, you will learn more specifics about the different plugins and how to use them.

How to Use Bootstrap JavaScript Plugins

The first step when using any of the Bootstrap plugins is to include the JavaScript file. There are three ways to include the JavaScript:

▶ Include the compiled, minified JavaScript file `<script src="js/bootstrap.min.js"> </script>`

▶ Include the compiled, non-minified JavaScript file `<script src="js/bootstrap.js"> </script>`

▶ Include only the plugin you want to use `<script src="js/dropdown.js"></script>`

Include the JavaScript *once* at the very bottom of your document, just above the `</html>` tag. Make sure that your `src` attribute points to the correct location and filename.

NOTE

Best Practice: Use the Minified Full Script File

Unless your website requires extreme download speeds, it's best to include the minified compiled JavaScript file, rather than including each plugin separately. This ensures that you always have the scripts ready if you add a new feature to the page, and the minified version is only 35KB, so it will still download fairly quickly.

After you have the Bootstrap JavaScript installed, you need to include jQuery as well. All the plugins require jQuery, so you should include a call to it just before the Bootstrap JavaScript. Bootstrap 3.3.2 requires jQuery 1.9.1 or higher. Listing 14.1 shows how the bottom of a typical Bootstrap document looks.

LISTING 14.1   The Bottom of a Bootstrap Document

```
<script src="http://code.jquery.com/jquery-latest.js"></script>
<script src="js/bootstrap.min.js"></script>
</body>
</html>
```

Setting Options for Plugins

Many of the plugins that Bootstrap offers include options so you can modify how the plugin works on your pages. The two ways you can include these options are

▶ Send the options as parameters in JavaScript method

▶ Use data attributes in the HTML

Both of these methods work equally well, and there are good reasons for using both in your websites. If you are comfortable writing JavaScript, you might find parameters easier, while data attributes require less knowledge of programming. Many Bootstrap sites use a combination of both to set the options.

Options as Parameters

If you're building the plugin completely in the JavaScript, this is a reasonable method to use. You trigger the plugin with JavaScript and include the options as a JSON array. Listing 14.2 shows how.

LISTING 14.2   Setting Options as Parameters

```
$('#example').tooltip({
  html:true,
  delay: 200,
  trigger:"click"
});
```

Listing 14.2 sets a tooltip (see Hour 17, "Pop-overs and Tooltips") with the options for HTML delivery, a delay of 200ms, and the trigger as a click. You also could put those options in an array variable and pass them that way.

Bootstrap plugins use JSON arrays to store the options data. If you want to learn more about JSON, you should visit the JSON website http://json.org/.

Options as Data Attributes

If the previous section with arrays, variables, JSON, and so on makes you nervous, you can relax because you don't need any of it to use Bootstrap plugins. Instead, you can use data attributes to assign the options to the plugins.

A data attribute is a new feature of HTML5. They are attributes on HTML elements that start with the word data- and then are followed by the name of the data attribute.

CAUTION

Data Attributes May Not Validate

If you use HTML validators to check your HTML, you may get error messages saying that the data-* attributes you use for Bootstrap plugins are not valid. Be sure that your web page uses the HTML5 doctype <!doctype html>; if you still get validation errors specific to the data-* attributes, you should ignore them or consider using a different HTML validator.

For example, in the Listing 14.2 tooltip, there was an option called delay. To set this option with data attributes, you would add an attribute to the tooltip element called data-delay. Listing 14.3 shows a button that includes a tooltip with the same options as in Listing 14.2, but the options are set with data attributes.

LISTING 14.3   Options Set with Data Attributes

```
<button type="button" class="btn btn-default" data-toggle="tooltip"
        data-html="true" data-delay="200" data-trigger="click"
        title="You did it!">
  Click to toggle Tooltip
</button>
```

As shown in Figure 14.1, the tooltip displays without any JSON or JavaScript variables or arrays.

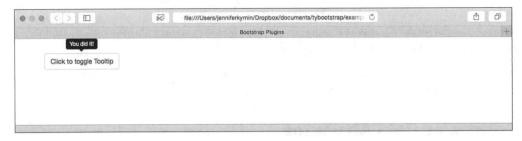

FIGURE 14.1
A tooltip using data attributes.

Data attributes should always be your first consideration when using Bootstrap plugins. They are easy to use, and placing the options right on the HTML element that defines the plugins makes the page easier to maintain. To look at the button example in Listing 14.2; it's obvious that a tooltip is assigned to this button because of the `data-toggle="tooltip"` attribute. As you look closer at the element, you know that the tooltip will have HTML results, have a delay of 200ms, and be triggered by a click. And if you want to change those options, you can do so right in the HTML.

But there are some problems with data attributes, including these:

▶ You can have only one plugin per element. In other words, your button cannot trigger both a tooltip and a modal. If you need to do that, you will need to wrap an element around the first element and attach the second plugin to that.

▶ You can set the options for only one element at a time. If you want to set every tooltip to display in HTML with a 200ms delay, you must add those data attributes to every element that has a tooltip.

▶ You cannot use complex JSON options in data attributes. For example, you might want to set the show delay on the tooltip to 200ms but the hide delay to 500ms. With JavaScript, you would set your option to `delay: { "show": 200, "hide": 500 }`. But there is no way to do that in Bootstrap. If you need to do that, you have to modify the `bootstrap.js` file.

If your website needs to have the data attributes disabled, you can do that as well. You add `$(document).off('.data-api')` to your document scripts, as shown in Listing 14.4.

LISTING 14.4 Turn Off Data Attributes

```
...
<script src="http://code.jquery.com/jquery-latest.js"></script>
<script src="js/bootstrap.min.js"></script>
<script>
  $(document).off('.data-api')
</script>
</body>
</html>
```

You also can turn off data attributes for specific plugins by including that plugin's name as a namespace along with the `data-api` namespace. For instance, to turn off data attributes for tooltips, you would write `$(document).off('.tooltip.data-api')`.

Using the JavaScript API

Bootstrap is designed to also use the JavaScript plugins through the JavaScript API. They have set up all the APIs as single, chainable methods that return the collection acted upon.

This means you can access the plugins from within your scripts and add them to other elements as you need them. For example, Listing 14.5 shows how you might add a tooltip to a button with an ID of #myToolTip.

LISTING 14.5   Add a Tooltip Programmatically

```
$('#myToolTip').tooltip('toggle');
```

All the Bootstrap methods accept three values:

▶ **Nothing**— For example, `tooltip()`. This indicates that the method should be initialized with the defaults.

▶ **An options object**— For example, `tooltip( { html: true } )`. This acts just like data attributes for setting the option values.

▶ **A string targeting a particular method**— For example, `tooltip('toggle')`. This initializes the tooltip with the `toggle` method invoked immediately.

You learn more about the specific options for each plugin in the following hours.

If you need access to the raw constructor for the plugin, it is available in the `Constructor` property—for example, `$.fn.tooltip.Constructor`. You also can use the `Constructor` property to change the default settings for a plugin. Listing 14.6 shows how you modify the `Constructor.DEFAULTS` object to change the default value.

LISTING 14.6   Change Tooltips to Default to HTML True

```
$.fn.tooltip.Constructor.DEFAULTS.html = true;
```

You can get the version number of the plugin with the `Constructor.VERSION` object—for example, `$.fn.tooltip.Constructor.VERSION`.

Events

Bootstrap provides custom events for most of the plugins. These are named in the infinitive (show) and past participle (shown) forms. The infinitive is triggered at the start of an event, and the past participle is triggered when the action is completed.

All Bootstrap events are namespaced as of version 3.0.0.

You also can stop an event before it starts with the preventDefault functionality. Listing 14.7 shows how you might do this.

LISTING 14.7 Preventing an Event with preventDefault

```
$('#myModal').on('show.bs.modal', function (e) {
  if (!data) return e.preventDefault() // stops modal showing
})
```

No Conflict

Sometimes you might want to use Bootstrap with another UI framework. This can cause conflicts if both Bootstrap and the other framework use the same names for things. To solve this, you can call .noConflict on the plugin you want to revert the value of. Then you can reassign the values to a nonconflicting name. Listing 14.8 shows how.

LISTING 14.8 Using .noConflict on a Button

```
// return $.fn.button to previously assigned value
var bootstrapButton = $.fn.button.noConflict();
// give $().bootstrapBtn the Bootstrap functionality
$.fn.bootstrapBtn = bootstrapButton;
```

CAUTION

Bootstrap Does Not Support Third-Party Libraries

Although Bootstrap has the .noConflict method and namespaced events, Bootstrap does not guarantee that these will work with third-party libraries and other frameworks. If you must use an additional library such as jQuery UI, you will have to test it and fix any problems you find on your own.

Disabled JavaScript

On very rare occasions your customers might disable JavaScript. Bootstrap is designed to work with JavaScript, and the plugins will not work if JavaScript is turned off.

When a customer with JavaScript turned off visits a page with a Bootstrap plugin, she might get strange results or see nothing at all. If you need to, you can use the <noscript></noscript> element to provide additional information, such as how to enable JavaScript. You also can create your own custom fallbacks for when JavaScript is disabled or turned off.

Summary

In this hour, you learned about how Bootstrap JavaScript plugins work. You learned what you need to do to install and use the plugins as well as how to set options for the plugins in several different ways.

This hour covered how to use the JavaScript API, including changing the default values, getting the plugin version, and triggering plugin events. You also learned about two possible problems—conflicts with other frameworks and disabled JavaScript—and how to solve them.

Workshop

The workshop contains quiz questions to help you process what you've learned in this hour. Try to answer all the questions before you read the answers.

Q&A

Q. What if I don't want to use any plugins on my pages?

A. If you don't use any plugins, then you don't need to include the Bootstrap JavaScript files or jQuery in your HTML. But by including them, you prepare your pages for the future if you do decide to add a plugin.

Q. I'm concerned about speed of downloading. Don't these plugins make the pages take longer?

A. Anything you add to your page is going to incrementally increase the download time. JavaScript, because it cannot load in parallel with any other element, can increase the time even more. But there are many things you can do to decrease this load time:

▶ **Use the minified JavaScript files**—The full file is only 35KB.

▶ **Use only the plugin scripts you need**—You will learn more about how to do this in Hour 21, "Customizing Bootstrap and Your Bootstrap Website."

▶ **Always place the JavaScript at the bottom of your documents**—This ensures that the rest of the page has loaded before the scripts start to download, thus making the page visible faster for your readers.

▶ **Use cached copies of the scripts**—By loading the script files from a content delivery network (CDN), you take advantage of web caches and help your files load more quickly. In the code samples in this hour, jQuery is installed from a CDN. You can even load Bootstrap from a CDN, such as the Microsoft Ajax Content Delivery Network (http://www.asp.net/ajax/cdn#Bootstrap_Releases_on_the_CDN_14).

Quiz

1. Which of these is not a way to include Bootstrap plugins?

    a. Include the full `bootstrap.js` file.

    b. Include the minified script file.

    c. Include the plugin JavaScript file.

    d. None of the above.

2. True or False: jQuery is an optional tool for using the Bootstrap plugins.

3. What is the best way to add options to plugins?

    a. As parameters

    b. As data attributes

    c. As script variables

    d. However works best for your page and your skills

4. How are parameters written?

    a. As a variable

    b. As an array

    c. As a JSON array

    d. As plain text

5. How do you add a data attribute?

    a. Add the tag `<data>` surrounding the information.

    b. Add an attribute called `data-*` with the name of the data (that is, `data-show`) to the HTML tag.

    c. Include the data in a `meta-data` attribute.

    d. Add an attribute with the name of the data (that is, `show`) to the HTML tag.

6. In this HTML, what is setting the option to open the tooltip after a lag?

```
<button type="button" class="btn btn-default" data-toggle="tooltip"
data-html="true" data-delay="200" data-trigger="click"
title="You did it!">
```

    a. `data-toggle="tooltip"`

    b. `data-delay="200"`

    c. `data-trigger="click"`

    d. `title="You did it!"`

7. Is this a valid way to add a tooltip?

```
$('#myToolTip').tooltip();
```

   a. Yes

   b. No

8. What does this line of code do?

```
$.fn.tooltip.Constructor.DEFAULTS.placement = 'bottom';
```

   a. Verifies that the tooltip `placement` default option is `bottom`

   b. Changes the tooltip `placement` default option to `bottom`

   c. Creates a new tooltip option called `placement` with a default of `bottom`

   d. Returns any tooltip options that are given the default value `bottom`

9. How do you return the version number of the dropdown plugin?

   a. `$.fn.tooltip.Constructor.VERSION`

   b. `$.fn.dropdown.constructor.VERSION`

   c. `$.fn.dropdown.Constructor.VERSION`

   d. `$.fn.dropdown.Constructor.Version`

10. How does Bootstrap handle a customer that doesn't have JavaScript?

   a. Bootstrap has an automatic fallback telling the customer to turn on JavaScript.

   b. Bootstrap has specific fallback options that are different for each plugin.

   c. Bootstrap forces the browser to turn on JavaScript.

   d. Bootstrap does nothing.

Quiz Answers

1. d. All of the answers are correct ways to include Bootstrap plugins.

2. False. You must include jQuery to use Bootstrap plugins.

3. d. There is no one best way to include options for Bootstrap plugins. You should use the method that works best for your scripts, your website, and your skills.

4. c. As a JSON array.

5. b. Add an attribute called `data-*` with the name of the data (that is, `data-show`) to the HTML tag.

6. b. `data-delay="200"`

7. a. Yes, it uses the Bootstrap API to add the tooltip directly to an element in the DOM.

8. b. Changes the tooltip `placement` default option to `bottom`.

9. c. `$.fn.dropdown.Constructor.VERSION`

10. d. Bootstrap does nothing. JavaScript is required to run Bootstrap plugins, but if it's not enabled, there is no fallback option. You have to build fallbacks yourself.

Exercises

Start considering what interactivity you are going to want on your website. Bootstrap offers features like tooltips, modals, dropdown menus, alerts, pop-overs, and much more. If you are not familiar with jQuery, you should consider learning how it is structured. A good book to start with is *Sams Teach Yourself JavaScript and jQuery in 24 Hours* by Brad Dayley.

Modal Windows

What You'll Learn in This Hour:

▶ How to build a modal window in Bootstrap
▶ Two ways to trigger a modal window
▶ How to adjust the size and layout of a modal
▶ Ways to use modals with dynamic content

In this hour you learn how to add modal windows to your Bootstrap pages. You also learn how to adjust the size and animation of modal windows and change the content in the modal windows and get some ideas for how you might use modals in your web designs.

What Is a Modal Window?

Modal windows force interactions by the user with the program at specific times. In most cases, they open and block access to the main window until the modal is taken care of. Whenever a program pops up a window that asks "Are you sure you want to delete that?" that window is a modal window. The user is expected to indicate whether or not the file should be deleted and cannot move on until he does so.

Modal windows on web pages might not be able to completely block interaction with the main web page, but otherwise they can act just like software program modals. They pop up a smaller window and allow the user to make changes without loading an entire new page.

You are not limited to asking questions in modals. You can use modal windows for lots of different things, including

▶ A lightbox for videos and images or a slideshow

▶ Contact and login forms

▶ Alerts and informative messages

- ▶ Search and results boxes

- ▶ Embedded media like PDFs or other documents

- ▶ Help windows

Bootstrap lets you easily create simple and complex modals on your web pages. Bootstrap modal windows are created as small windows with rounded corners that are displayed in a lightbox effect over the top of the regular content. The regular content is then faintly grayed out to give the modal more emphasis. You can see an example of a Bootstrap modal window in Figure 15.1.

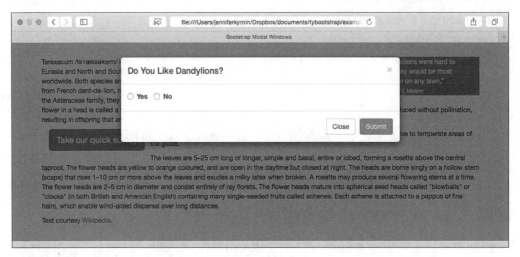

FIGURE 15.1
A standard Bootstrap modal window.

How to Build a Modal Window

You need two things to add a modal window to your Bootstrap pages: the HTML for the window and the HTML or JavaScript for the trigger to open the modal window. Most people find it easiest to build the trigger first, so that when they build the window, they can turn it on and see it.

But before you do either, you should make sure that you have jQuery and the Bootstrap JavaScript file loaded at the bottom of your web document. You can use either `bootstrap.js` (or the minified version) or `modal.js` if you plan to use only that one plugin.

Triggering a Modal

As you learned in Hour 14, "Using Bootstrap JavaScript Plugins," you can trigger the modal plugin in two ways: with JavaScript or with a data attribute.

Listing 15.1 shows how to trigger a modal with JavaScript.

LISTING 15.1   JavaScript to Trigger a Modal

```
$('#myModal').modal(options);
```

This opens a modal with the ID myModal with the options in the options array.

The three options you can set on your modals are

▶ backdrop—This tells the browser whether there should be a backdrop (graying out the original page content). The default is true. Setting it to false removes the backdrop completely and forces the user to click a close button on the modal. Additionally, a value of static leaves the backdrop but removes the ability to close the modal except with the close buttons on the modal.

▶ keyboard—If this option is true, the modal can be closed with the escape key. If it's false, the escape key won't close the window.

▶ show—If this option is true, the modal will be shown when it is initialized. If it is false, the modal will be initialized but not displayed.

You might see a deprecated option, remote, on older versions of Bootstrap. This was used to load a modal from a separate HTML document on your server. It will be removed in Bootstrap 4, so you should not use it. Instead, you should use client-side templating or a data binding framework, or call jQuery load() yourself.

The other way to trigger a modal is with data attributes. You can trigger a modal by including the attribute data-toggle="modal". You need to include a reference to the modal being displayed with either data-target="#myModal" on <button> tags or href="#myModal" on <a> tags. Listing 15.2 shows a button that triggers a modal when it's clicked.

LISTING 15.2   A Button to Trigger a Modal

```
<button type="button" class="btn btn-warning btn-lg"
        data-toggle="modal" data-target="#myModal">
  Click to Open a Modal
</button>
```

Listing 15.3 shows that same modal being triggered by a link.

LISTING 15.3   A Link to Trigger a Modal

```
<a class="bg-warning" data-toggle="modal" href="#myModal">
  Click to Open a Modal
</a>
```

NOTE

Bootstrap Adds Classes When Modals Are Triggered

After a modal is triggered, Bootstrap adds the `.modal-open` class to the `<body>` tag. It also adds an empty `<div>` with the class `.modal-backdrop` to the page. These are required to create the modal and the backdrop, if there is one. However, you can use them in your style sheet as well.

The JavaScript for modals includes several methods you can use in your scripts:

▶ `toggle`—This manually toggles the modal on and off. It returns to the handler before the modal is actually shown or hidden.

▶ `show`—This manually opens the modal. It returns to the caller before the modal is shown.

▶ `hide`—This manually closes the modal. It returns to the caller before the modal is hidden.

▶ `handleUpdate`—This readjusts the modal's positioning to counter a scrollbar should one appear. Without this method, if a scrollbar appears, the modal jumps to the left. It is needed only if the modal height changes while it's open.

In addition, several events are available that you can use to hook into the functionality in your scripts:

▶ `show.bs.modal`—This fires as the `show` method is called. If this is caused by a click, the clicked element is available as the `relatedTarget` property of the event.

▶ `shown.bs.modal`—This event fires when the modal has been made visible to the user (after any CSS transitions). If this is caused by a click, the clicked element is available as the `relatedTarget` property of the event.

▶ `hide.bs.modal`—This event fires immediately when the `hide` method is called.

▶ `hidden.bs.modal`—This event fires when the modal has been made completely hidden from the user (including any CSS transitions).

▶ `loaded.bs.modal`—This event fires when the modal has loaded content using the `remote` option. This has been deprecated in Bootstrap 3.3.0.

Coding a Modal

After you have a trigger, you need to create the modal window itself. This is just a block of HTML that is given modal classes. First, you need a container element with the `modal` class and the ID you referenced in the trigger. This contains the entire modal window:

```
<div class="modal" id="myModal">
```

If you place your modal content in this container, your modal will not have a white background color and will be the full width of the screen. You need two more containers to get your modal window to look correct:

- ▶ .modal-dialog—This sets the width of the modal window.

- ▶ .modal-content—This defines the content area and sets the background color for the modal window.

Listing 15.4 shows the minimum HTML you need for a reasonable looking modal window.

LISTING 15.4   The Minimum HTML for a Modal

```
<div class="modal" id="myModal">
  <div class="modal-dialog">
    <div class="modal-content">
      <p>My modal content</p>
    </div>
  </div>
</div>
```

Inside the .modal-content container you can have several additional classes to define various parts of the modal:

- ▶ .modal-header—This defines the header or top of the modal. In the header you should include a close element so that your modal can be closed. You also can include a title.

- ▶ .modal-title—This is the title of the modal.

- ▶ .modal-body—The main content of the modal.

- ▶ .modal-footer—This is the footer or bottom part of the modal window. Most modals include submit and cancel buttons in this area.

Inside your modal you always should include a close button; many include two—one at the top and one at the bottom. For the top close button, you can create a small *x* using the .close class you learned about in Hour 13, "Bootstrap Utilities." You also should include the data attribute data-dismiss with the value modal. This tells Bootstrap that when the close button is clicked, the modal should be dismissed:

```
<button type="button" class="close" data-dismiss="modal"
        aria-label="Close">
  <span aria-hidden="true">&times;</span>
</button>
```

If you want your close button to look like a standard button, you don't need the `.close` class; you do still need the `data-dismiss` attribute, though:

```
<button type="button" class="btn btn-default" data-dismiss="modal">
  Close
</button>
```

Listing 15.5 shows the HTML for a fancy modal window including both types of close button.

LISTING 15.5   A Fancy Modal Window

```
<div class="modal" id="myModal">
  <div class="modal-dialog">
    <div class="modal-content">
      <div class="modal-header">
        <button type="button" class="close" data-dismiss="modal"
                aria-label="Close">
          <span aria-hidden="true">&times;</span>
        </button>
        <h4 class="modal-title">Do You Love Dandylions?</h4>
      </div>
      <div class="modal-body">
        <img src="images/seeded.jpg" alt=""
             class="col-sm-3 img-circle pull-left"/>
        <p>So do I!</p>
      </div>
      <div class="modal-footer">
        <button type="button" class="btn btn-default"
                data-dismiss="modal">Close</button>
      </div>
    </div>
  </div>
</div>
```

▼ TRY IT YOURSELF

Trigger a Modal When the Page Loads

One popular technique with modal windows is to have one load after the page loads to provide a greeting, give new customers additional information, or promote something. This is easy to do with just a few lines of JavaScript along with your Bootstrap code:

1. Open the page where you want the modal to load in an HTML editor.

2. Confirm that you have jQuery and the Bootstrap JavaScript loaded at the bottom of the page:

    ```
    <script src="http://code.jquery.com/jquery-latest.js"></script>
    <script src="js/bootstrap.min.js"></script>
    ```

3. Add the HTML for your modal window to the HTML. Give it an ID of #myModal:

```
<div class="modal fade" id="myModal">
...
</div>
```

4. Below the Bootstrap JavaScript and jQuery, add a <script> element:

```
<script>
</script>
```

5. Inside the <script> tags, include a document ready function:

```
$(document).ready(function(){
});
```

6. Inside the document ready function, show your modal window on the #myModal element:

```
$('#myModal').modal('show');
```

This will load the modal immediately after the page loads. If you want to have the modal appear a few seconds later, you'll need to add a timeout to it. Figure 15.2 shows how the modal would look, and Listing 15.6 shows how to build it in a Bootstrap page.

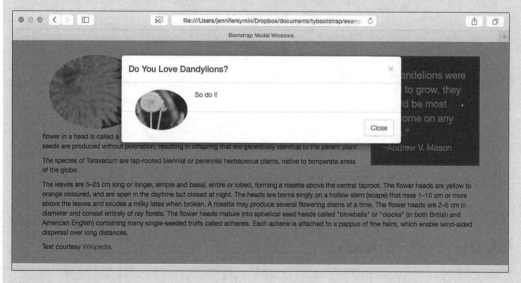

FIGURE 15.2
A modal that loads automatically.

▼ **LISTING 15.6** Trigger a Modal to Load Two Seconds After the Page Loads

```
<div class="modal fade" id="myModal">
  <div class="modal-dialog">
    <div class="modal-content">
      <div class="modal-header">
        <button type="button" class="close" data-dismiss="modal"
                aria-label="Close">
          <span aria-hidden="true">&times;</span>
        </button>
        <h4 class="modal-title">Do You Love Dandylions?</h4>
      </div>
      <div class="modal-body">
        <img src="images/seeded.jpg" alt=""
             class="col-sm-3 img-circle pull-left"/>
        <p>So do I!</p>
      </div>
      <div class="modal-footer">
        <button type="button" class="btn btn-default"
                data-dismiss="modal">Close</button>
      </div>
    </div>
  </div>
</div>

<script src="http://code.jquery.com/jquery-latest.js"></script>
<script src="js/bootstrap.min.js"></script>
<script>
$(document).ready(function(){
  function show_modal() {
    $('#myModal').modal('show');
  }
  window.setTimeout(show_modal, 2000);
});
</script>
```

Modifying Modals

A few things you can do with Bootstrap to modify how your modals look and act on the page include changing how they open on the page, the size, the layout inside them, and even the content based on how the user triggered the modal.

Changing How a Modal Opens

The default way for a modal to open is just to blink onto the screen. There is no transition—just click and it's there. But this can be jarring, so most modals include the class .fade so that they fade in more gracefully:

```
<div class="modal fade" id="myModal">
```

Comparing Fade and No Fade Modals

The best way to understand the difference between modals that fade in and ones that just appear is to do it. This Try It Yourself walks you through creating two modals—one that fades in and one that just appears:

1. Open a web page, and add the Bootstrap CSS and JavaScript in the standard template.

2. Create two trigger buttons. One will target a #fade ID; the other will target a #noFade ID:

   ```
   <button type="button" class="btn btn-default" data-toggle="modal"
           data-target="#noFade">
     No Fade Modal
   </button>
   <button type="button" class="btn btn-default" data-toggle="modal"
           data-target="#fade">
     Fade Modal
   </button>
   ```

3. Create two modals, one with the ID #fade and one with the ID #noFade:

   ```
   <div class="modal" id="noFade" tabindex="-1" role="dialog"
        aria-labelledby="noFadeLabel" aria-hidden="true">
     <div class="modal-dialog">
       <div class="modal-content">
         ...
       </div>
     </div>
   </div>
   ```

4. Add the fade class to the #fade modal:

   ```
   <div class="modal fade" id="fade" tabindex="-1" role="dialog"
        aria-labelledby="fadeLabel" aria-hidden="true">
   ```

5. Save the page, and open it in your web browser to try them. Listing 15.7 shows my final HTML.

▼ **LISTING 15.7**   Comparing Fading and Nonfading Modals

```
<!DOCTYPE html>
<html lang="en">
  <head>
    <meta charset="utf-8">
    <meta http-equiv="X-UA-Compatible" content="IE=edge">
    <meta name="viewport" content="width=device-width, initial-scale=1">
    <title>Bootstrap Modal Windows</title>

    <!-- Bootstrap -->
    <link href="css/bootstrap.min.css" rel="stylesheet">
    <!-- HTML5 shim and Respond.js for IE8 support of HTML5
    elements and media queries -->
    <!-- WARNING: Respond.js doesn't work if you view the page via
    file:// -->
    <!--[if lt IE 9]>
      <script
src="https://oss.maxcdn.com/html5shiv/3.7.2/html5shiv.min.js">
      </script>
      <script
src="https://oss.maxcdn.com/respond/1.4.2/respond.min.js"></script>
    <![endif]-->

  </head>
  <body>
    <p> </p>
    <div class="container">

      <button type="button" class="btn btn-default"
            data-toggle="modal" data-target="#noFade">
        No Fade Modal
      </button>
      <button type="button" class="btn btn-default"
            data-toggle="modal" data-target="#fade">
        Fade Modal
      </button>


<!-- Modal -->
<div class="modal" id="noFade" tabindex="-1" role="dialog"
     aria-labelledby="noFadeLabel" aria-hidden="true">
  <div class="modal-dialog">
    <div class="modal-content">
      <div class="modal-header">
        <button type="button" class="close" data-dismiss="modal"
                aria-label="Close">
```

```
          <span aria-hidden="true">&times;</span>
        </button>
        <h4 class="modal-title" id="noFadeLabel">This Modal Did Not
        Fade In</h4>
      </div>
      <div class="modal-body">
        <p>It just blinked into existance. Poof!</p>
      </div>
    </div>
  </div>
</div>
<div class="modal fade" id="fade" tabindex="-1" role="dialog"
    aria-labelledby="fadeLabel" aria-hidden="true">
  <div class="modal-dialog">
    <div class="modal-content">
      <div class="modal-header">
        <button type="button" class="close" data-dismiss="modal"
                aria-label="Close">
          <span aria-hidden="true">&times;</span>
        </button>
        <h4 class="modal-title" id="noFadeLabel">This Modal Did
        Fade In</h4>
      </div>
      <div class="modal-body">
        <p>It appeared to slide onto the page. Whee!</p>
      </div>
    </div>
  </div>
</div>
    </div>

<script src="http://code.jquery.com/jquery-latest.js"></script>
<script src="js/bootstrap.min.js"></script>

</body>
</html>
```

Changing the Size of Modals

There are three sizes for modals: small, medium, and large. The default is medium. The classes
.modal-lg and .modal-sm are placed on the .modal-dialog container and change the size to
large and small, respectively. Figure 15.3 shows the three sizes of modals, and Listing 15.7 shows
the HTML for a large modal.

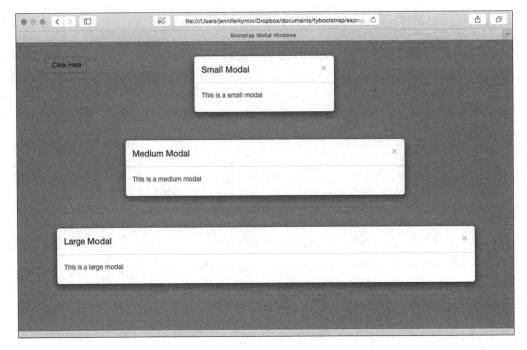

FIGURE 15.3
Three modal sizes: small, medium, and large.

Changing the Layout

You can use the Bootstrap grid system within the modal windows to adjust the layout inside the window. All you need to do is add a `.container-fluid` inside the `.modal-body` and then add grid elements, just like you learned in Hour 5, "Grids and How to Use Them."

▼ TRY IT YOURSELF

Create a Four-Column Modal

With the Bootstrap grids, you easily can create a modal that has four columns rather than just one. This Try It Yourself walks you through how to create a four-column modal window:

1. Open your Bootstrap page in an HTML editor.

2. Add the modal trigger button:

   ```
   <button type="button" class="btn btn-default" data-toggle="modal"
           data-target="#myModal">
     Click Here
   </button>
   ```

3. Create the modal window:

```
<!-- Modal -->
<div class="modal" id="myModal" tabindex="-1" role="dialog"
     aria-labelledby="myModalLabel" aria-hidden="true">
  <div class="modal-dialog modal-lg">
    <div class="modal-content">
      <div class="modal-header">
        <button type="button" class="close" data-dismiss="modal"
                aria-label="Close">
          <span aria-hidden="true">&times;</span>
        </button>
        <h4 class="modal-title" id="myModalLabel">My Modal</h4>
      </div>
      <div class="modal-body">
        <p>This is a four-column modal</p>
        <p>Each of these paragraphs is in a different column.</p>
        <p>This is column three.</p>
        <p>And this is the final column, column four.</p>
      </div>
    </div>
  </div>
</div>
```

4. Inside the `.modal-body` container, add a `<div class="container-fluid">` tag with a `<div class="row">` inside that.

5. Make sure the four paragraphs are inside the `.row` container, and then add the `.col-md-3` class to each of them:

```
<p class="col-md-3">This is a four-column modal</p>
<p class="col-md-3">Each of these paragraphs is in a different
column.</p>
<p class="col-md-3">This is column three.</p>
<p class="col-md-3">And this is the final column, column four.</p>
```

6. Save the page, and open it in a web browser to test. Listing 15.8 has the full HTML for this page.

LISTING 15.8 A Four-Column Modal

```
<!DOCTYPE html>
<html lang="en">
  <head>
    <meta charset="utf-8">
    <meta http-equiv="X-UA-Compatible" content="IE=edge">
    <meta name="viewport" content="width=device-width, initial-scale=1">
    <title>Bootstrap Modal Windows</title>
```

```
    <!-- Bootstrap -->
    <link href="css/bootstrap.min.css" rel="stylesheet">
    <!-- HTML5 shim and Respond.js for IE8 support of HTML5
    elements and media queries -->
    <!-- WARNING: Respond.js doesn't work if you view the page via
    file:// -->
    <!--[if lt IE 9]>
      <script
src="https://oss.maxcdn.com/html5shiv/3.7.2/html5shiv.min.js">
      </script>
      <script
src="https://oss.maxcdn.com/respond/1.4.2/respond.min.js"></script>
    <![endif]-->

  </head>
  <body>
    <p> </p>
    <div class="container">

      <button type="button" class="btn btn-default"
              data-toggle="modal" data-target="#myModal">
        Click Here
      </button>

<!-- Modal -->
<div class="modal" id="myModal" tabindex="-1" role="dialog"
     aria-labelledby="myModalLabel" aria-hidden="true">
  <div class="modal-dialog modal-lg">
    <div class="modal-content">
      <div class="modal-header">
        <button type="button" class="close" data-dismiss="modal"
                aria-label="Close">
          <span aria-hidden="true">&times;</span>
        </button>
        <h4 class="modal-title" id="myModalLabel">My Modal</h4>
      </div>
      <div class="modal-body">
        <div class="container-fluid">
          <div class="row">
            <p class="col-md-3">This is a four-column modal</p>
            <p class="col-md-3">Each of these paragraphs is in a
            different column.</p>
            <p class="col-md-3">This is column three.</p>
            <p class="col-md-3">And this is the final column,
            column four.</p>
          </div>
        </div>
```

```
        </div>
      </div>
    </div>
  </div>
      </div>

<script src="http://code.jquery.com/jquery-latest.js"></script>
<script src="js/bootstrap.min.js"></script>

</body>
</html>
```

Changing Modal Content Dynamically

You can create a dynamic page that displays different content inside a modal depending on which trigger is selected. This uses the event.relatedTarget and data-* attributes to change the contents of the modal when different triggers are clicked. Listing 15.9 shows HTML and JavaScript that create a photo gallery that opens large size images when the thumbnails are clicked.

LISTING 15.9   A Modal Photo Gallery

```
<a href="#" data-toggle="modal" data-target="#myModal"
   data-imagetitle="Shasta and McKinley"
   data-imagesource="images/pic1.png">
  <img src="images/thumb1.png" alt="shasta and mckinley"
       class="img-thumbnail">
</a>
<a href="#" data-toggle="modal" data-target="#myModal"
   data-imagetitle="Shasta" data-imagesource="images/pic2.png">
  <img src="images/thumb2.png" alt="shasta" class="img-thumbnail">
</a>

<div class="modal fade" id="myModal" tabindex="-1" role="dialog"
     aria-labelledby="myModalLabel" aria-hidden="true">
  <div class="modal-dialog">
    <div class="modal-content">
      <div class="modal-header">
        <button type="button" class="close" data-dismiss="modal"
                aria-label="Close">
          <span aria-hidden="true">&times;</span>
        </button>
        <h4 class="modal-title" id="myModalLabel">Image</h4>
      </div>
      <div class="modal-body">
```

```
      </div>
    </div>
  </div>
</div>

...

<script>
$('#myModal').on('show.bs.modal', function (event) {
  var button = $(event.relatedTarget);
  var recipient = button.data('imagetitle');
  var source = button.data('imagesource');
  var modal = $(this);
  modal.find('.modal-title').text(recipient);
  modal.find('.modal-body').html('<img src="' + source + '" alt="'
+ recipient + '" class="center-block">');
})
</script>
```

Summary

This hour covered the Bootstrap plugin modal.js. You learned how to create and use modal windows to provide more information to customers in windows that open above the existing content. You also learned several ways you can use modal windows, how to trigger them, and how to style them.

There are a lot of options, methods, events, and CSS classes you can use to work with modals. Tables 15.1–15.4 detail the options, methods, events, and classes covered in this hour.

TABLE 15.1   Modal Options

Option	Value	Description
backdrop	true or false or the string 'static'	Indicates whether there should be a backdrop (true) or not (false). The static keyword indicates that there should be a backdrop, but it won't close the modal when it's clicked. The default is true.
keyboard	true or false	Allows the modal to be closed when the escape key is pressed. The default is true.
remote	URL path	**This option is deprecated as of Bootstrap 3.3.0 and should not be used.** This indicates content that will be loaded into the .modal-content container. The default is false, and no content will be loaded.
show	true or false	Shows the modal when initialized. The default is true.

TABLE 15.2 Modal Methods

Method	Description
`.modal(options)`	Activates your content as a modal with the optional `options` array.
`.modal('handleUpdate')`	If the modal's height changes while it's open, this method readjusts the positioning to counter a scrollbar and prevents the modal from jumping on the screen.
`.modal('hide')`	Manually hides the modal.
`.modal('show')`	Manually shows the modal.
`.modal('toggle')`	Manually toggles the modal on or off.

TABLE 15.3 Modal Events

Event	Description
`hidden.bs.modal`	Fired after the modal is completely hidden from the user.
`hide.bs.modal`	Fired as the modal is hidden.
`loaded.bs.modal`	Fired when the modal has loaded content with the `remote` option.
`show.bs.modal`	Fired as the modal is displayed.
`shown.bs.modal`	Fired after the modal is completely displayed to the user.

TABLE 15.4 Modal CSS Classes

Class	Description
`fade`	Adds an animation to the loading modal to make it appear to fade in when it loads.
`modal`	Indicates that the element is a modal.
`modal-backdrop`	Automatically placed on a `<div>` to set the backdrop.
`modal-body`	Indicates that the element is the body of the modal.
`modal-content`	Indicates that the element is the main content of the modal.
`modal-dialog`	Defines the dialog box of the modal. This also defines the width of the modal.
`modal-footer`	Indicates that the element is the footer of the modal.
`modal-header`	Indicates that the element is the header of the modal.
`modal-lg`	Makes a large modal.
`modal-open`	Automatically placed on the `<body>` element when a modal is open on the screen.
`modal-sm`	Makes a small modal.
`modal-title`	Indicates the element is the title of the modal.

Workshop

The workshop contains quiz questions to help you process what you've learned in this hour. Try to answer all the questions before you read the answers.

Q&A

Q. Isn't a modal just a lightbox?

A. Not exactly. A lightbox is a type of modal that has been optimized to display images and other media, but modal windows can do more than just display images.

Q. What are the other attributes you list in the modals, such as `role="dialog"`, `aria-labelledby="myModalLabel"`, **and** `aria-hidden="true"`**?**

A. These attributes are used to make the modal more accessible. The `role` attribute indicates that the modal is a dialog box, and the two `aria-` attributes indicate the label and if it should be hidden. These and other accessibility features are covered in more detail in Hour 21, "Customizing Bootstrap and Your Bootstrap Website."

Q. I really want to use the `remote` **option to load an external web page. How do I use it?**

A. You can set it on your trigger button with the `data-remote` attribute. Set this to the URL you want to load. If your trigger is a link, you can use the `href` attribute to define the page to load—for example, `<a data-toggle="modal" href="remote.html" data-target="#modal">Click me</a>`. However, this doesn't work in all web browsers.

Quiz

1. Which of the following can be created with a Bootstrap modal?

   a. A login box

   b. A photo lightbox

   c. An alert

   d. B and C

   e. All of the above

2. Will this trigger a modal?

```
$('#myModal').modal({
    "backdrop" : "static"
});
```

   a. Yes, with a static backdrop

   b. Yes, with a standard backdrop

c. No, because the backdrop `static` value is incorrect

d. No, because you cannot include options that way

3. Will this trigger a modal?

```
<a class="bg-warning" data-toggle="modal">
  Click to Open a Modal
</a>
```

a. Yes, a standard modal

b. Yes, a modal with a warning background

c. No, because there is no target

d. No, because there is no URL

4. Which classes are required to create a modal window?

a. `.modal.`

b. `.modal` and `.modal-dialog.`

c. `.modal`, `.modal-dialog`, and `.modal-content.`

d. There are no required classes for modals.

5. Which of the following defines the title of a modal?

a. `<div class="title">Modal Title</div>`

b. `<h2 class="modal-title">Modal Title</h2>`

c. `<h1>Modal Title</h1>`

d. `Modal Title`

6. Why should you avoid using the `remote` option?

a. Because it's not a valid option.

b. Because it doesn't work on all browsers.

c. Because it's deprecated.

d. You don't need to avoid using it.

7. How do you animate the loading of a modal?

a. Use the `animate` class.

b. Use the `fade` class.

c. You don't do anything because they are animated by default.

d. You can't animate the loading of a modal.

8. Where do you put the `.modal-sm` class to create a small modal?

   a. On the `.modal` element

   b. On the `.modal-dialog` element

   c. On the `.modal-content` element

   d. On a new container element around the whole modal

9. True or False: You cannot use column grid classes inside modals.

10. Where do you put the `.modal-open` class?

   a. You place it on the `<body>` element to indicate that the page has a modal.

   b. You place it on the `.container` element to indicate there is a modal in the container.

   c. You don't put it anywhere. Bootstrap adds it automatically to the `.modal` element when the modal is opened.

   d. You don't put it anywhere. Bootstrap adds it automatically when the modal is opened.

Quiz Answers

1. e. All of the above.

2. a. Yes, with a static backdrop.

3. c. No, because there is no target.

4. c. `.modal`, `.modal-dialog`, and `.modal-content` are required to build correct-looking Bootstrap modals.

5. b. `<h2 class="modal-title">Modal Title</h2>`

6. c. Because it's deprecated.

7. b. Use the `fade` class.

8. b. On the `.modal-dialog` element.

9. False. You can use column grid classes inside modals.

10. d. You don't put it anywhere. Bootstrap adds it automatically when the modal is opened.

Exercise

Add a modal to one of your web pages. Trigger it with a link or a button.

Affix, Tab, and ScrollSpy

What You'll Learn in This Hour:

▶ How to fix items to the screen, so they don't scroll
▶ How to build tabbed content panes
▶ How to create navigation that changes as the page scrolls
▶ How to use the Affix, Tab, and ScrollSpy plugins together

Affix, Tab, and ScrollSpy are Bootstrap plugins that give you more flexibility with your navigation and other elements on the page. Like every other Bootstrap plugin, they can be triggered through JavaScript or right in the HTML with `data-*` attributes. In this hour you learn how to use these three plugins.

Affix

Affix is a plugin that lets you create a pinned element on your page. Many sites use this to keep the navigation visible on the page no matter where the customer scrolls. But you can also use it to pin images or media as well as blocks of text.

The Affix plugin uses the `affix.js` script to toggle `position:fixed;` on and off depending on the scroll position on the page. Some common uses for the Affix plugin include

▶ Internal page navigation
▶ Social media sharing buttons
▶ Advertising

The plugin works by setting a top position where the element will start on the page. After the user scrolls past the element, it stays fixed on the screen with the content scrolling beside it. In most situations, designers leave the element fixed for the entire height of the page, but with the Affix plugin you can specify a position where the element should stop being fixed and start scrolling with the page again.

Using Affix

To use the Affix plugin, you first must include both jQuery and the `affix.js` file at the bottom of the HTML document. If you include the full Bootstrap JavaScript file, Affix will be included.

Then you need to place the element to be fixed on the document. This can be as simple as just placing the HTML in the document, or you can use CSS to precisely position it.

Finally, add the `data-spy="affix"` attribute to the element. Test your page in a few browsers and devices. Remember that the page needs to be long enough to scroll so the Affix plugin will take effect. Listing 16.1 shows the HTML for a navigation list affixed to the top of the page.

LISTING 16.1   Affix In-Page Navigation

```
<!DOCTYPE html>
<html lang="en">
  <head>
    <meta charset="utf-8">
    <meta http-equiv="X-UA-Compatible" content="IE=edge">
    <meta name="viewport"
          content="width=device-width, initial-scale=1">
    <title>Bootstrap Modal Windows</title>

    <!-- Bootstrap -->
    <link href="css/bootstrap.min.css" rel="stylesheet">
    <!-- HTML5 shim and Respond.js for IE8 support of HTML5
    elements and media queries -->
    <!-- WARNING: Respond.js doesn't work if you view the page via
    file:// -->
    <!--[if lt IE 9]>
      <script
src="https://oss.maxcdn.com/html5shiv/3.7.2/html5shiv.min.js">
      </script>
      <script
src="https://oss.maxcdn.com/respond/1.4.2/respond.min.js"></script>
    <![endif]-->

  </head>
<body>

  <div class="container">
    <nav class="col-md-4">
      <ul class="list-group" data-spy="affix">
        <li class="list-group-item">
          <a href="#section1">Section One</a></li>
        <li class="list-group-item">
          <a href="#section2">Section Two</a></li>
```

```
          <li class="list-group-item">
             <a href="#section3">Section Three</a></li>
        </ul>
      </nav>
      <div class="col-md-8">
        <h2 id="section1">Section One</h2>
<p>Lorem ipsum dolor sit amet, consectetur adipiscing elit. ...</p>
        <h2 id="section2">Section Two</h2>
        <p>Suspendisse ornare ipsum nec velit euismod egestas. ...</p>
        <h2 id="section3">Section Three</h2>
        <p>Interdum et malesuada fames ac ante ipsum primis ...</p>
      </div>
  </div>

<script src="http://code.jquery.com/jquery-latest.js"></script>
<script src="js/bootstrap.min.js"></script>

</body>
</html>
```

The Affix plugin uses three CSS classes that represent a particular state:

▶ .affix-top— The Affix plugin places this on the element when it's in the topmost position.

▶ .affix— When the element is scrolled past, this triggers the .affix class to replace the .affix-top class and adds the style position: fixed; from the Bootstrap CSS.

▶ .affix-bottom— This class replaces the .affix class if an offset is defined.

There are two offsets you can use to define when the plugin should trigger (top) and when it should stop (bottom). You can define these with data attributes:

▶ data-offset-top="60"

▶ data-offset-bottom="120"

Or you can define them with JavaScript, as in Listing 16.2.

LISTING 16.2 Setting Offsets with JavaScript

```
$('#myAffix').affix({
  offset: {
    top: 60,
    bottom: 120
  }
});
```

The last option you can set is the `target` option. This defaults to the `window` object, but you can specify the selector, node, or jQuery object to be Affix if needed.

Several events exist in the Affix plugin:

▶ `affix.bs.affix`—This fires just before the element has been affixed.

▶ `affixed.bs.affix`—This fires right after the element has been affixed.

▶ `affix-top.bs.affix`—This fires immediately before the `.affix-top` class is added.

▶ `affixed-top.bs.affix`—This event fires after the element has had the `.affix-top` class applied.

▶ `affix-bottom.bs.affix`—This fires immediately before the `.affix-bottom` class is added.

▶ `affixed-bottom.bs.affix`—This event fires after the element has had the `.affix-bottom` class applied.

Tab

The Tab plugin lets you add "toggleable" tabs—tabs that can be toggled on or off—to show and hide panes of content. This extends the existing tabbed navigation you learned about in Hour 12, "Creating Navigation Systems with Bootstrap," to add toggleable areas. Figure 16.1 shows a basic tab area.

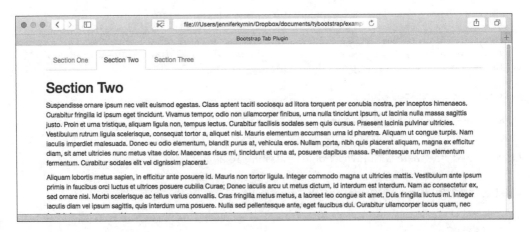

FIGURE 16.1
A basic tab area with three tabs.

Tab   261

Using Tab

As with every other plugin, you must include both jQuery and the Bootstrap JavaScript in your document. The Tab plugin also requires that you use a tabbed navigation component like you learned in Hour 12.

To add tabs, you need HTML for the tab bar. This is typically an unordered list with the `.nav` and `.nav-tabs` classes.

Then you need a `<div>` or other element that contains the tab information. Inside that element, place a separate `<div>` with a unique ID for each tab. Link the unordered list elements to the IDs of the tab information `<div>` elements. Listing 16.3 shows the HTML before you add the Tab plugin.

LISTING 16.3   HTML Before the Tab Plugin Is Added

```
<ul class="nav nav-tabs">
  <li class="active"><a href="#section1">Section One</a></li>
  <li><a href="#section2">Section Two</a></li>
  <li><a href="#section3">Section Three</a></li>
</ul>
<div>
  <div id="section1">
    <h2>Section One</h2>
    <p>...</p>
  </div>
  <div id="section2">
    <h2>Section Two</h2>
    <p>...</p>
  </div>
  <div id="section3">
    <h2>Section Three</h2>
    <p>...</p>
  </div>
</div>
```

NOTE

You Also Can Use Tab with `.nav-pills`

Although the Tab plugin implies that you can use only the `.nav-tabs` class, it works just fine with `.nav-pills` as well. Just be aware that the tabbed styling that Bootstrap gives to `.nav-tabs` lends itself to this type of content. Your customers are more likely to recognize that there is hidden content behind the tabs than if there are pill buttons at the top instead. It is also interesting to note that you don't need to use the `.nav` and `.nav-tabs/pills` classes at all. The Tab plugin will work with plain links just as well as with navigation tabs or buttons.

Add the class .tab-content to the element containing the tab content <div> elements. Each tab content div should have the .tab-pane class.

You can activate the Tab plugin with data attributes or with JavaScript. With JavaScript, you need to enable each tab individually. Listing 16.4 shows a simple script that will enable each tab when it is clicked on.

LISTING 16.4   Enable a Tab When It's Clicked On

```
$('#myTab a').click(function (e) {
  $(this).tab('show');
});
```

This script will show the ID of the tab in the URL as a #hash. If you don't want that to display, add the line e.preventDefault(); to the function above the $(this) line.

CAUTION

The URL Should Change, but You Have to Catch That

One frustration with dynamic content is that it is hard to bookmark the page in a different state than the default, such as with a different tab than the initial one open. If you include the e.preventDefault(); line in your script, you make the page less accessible because the URL does not change and your customers cannot bookmark the open tabs. However, if you remove this line, you have to write a script to load the tab pane called in the URL. Listing 16.5 shows a simple script you can use to make the hash tags in the URL open the correct tab.

LISTING 16.5   Make URLs Work to Open Different Panes

```
$(function(){
  var hash = window.location.hash;
  hash && $('ul.nav a[href="' + hash + '"]').tab('show');

  $('#myTab a').click(function (e) {
    $(this).tab('show');
  });
});
```

Tabs are activated when a customer clicks on them, but you also can activate them with JavaScript using the $().tab('show') method. You can use several different jQuery selectors to select specific tabs to activate. For example:

▶ $('#myTab a[href="#id"]').tab('show');—Select the tab by where it links to.

▶ $('#myTab a:**first**').tab('show');—Select the first tab.

▶ $('#myTab a:**last**').tab('show');—Select the last tab.

▶ $('#myTab **li:eq(2)** a').tab('show');—Select the third tab (starting tab is number 0).

Tab   263

Activate the Tab plugin with `data-*` attributes. Add the `data-toggle="tab"` attribute to the links in your tab list. If you are using pills instead of tabs, you can use `data-toggle="pill"` as well. But the style of the tabs is defined by the `.nav-tabs` or `.nav-pills` class on the list, not by which `data-toggle` you use.

Bootstrap enables you to use the `.fade` class (refer to Hour 15, "Modal Windows") to have the tab panel contents appear more dynamically. Just add the `.fade` class to the `.tab-pane` elements. You also need to add the `.in` class on the active pane so it shows up immediately. Listing 16.6 shows the HTML for this effect.

LISTING 16.6 Add a Fade Effect

```
<div class="tab-content">
  <div role="tabpanel"
      class="tab-pane fade in active" id="section1">
    <h2>Section One</h2>
    <p>...</p>
  </div>
  <div role="tabpanel" class="tab-pane fade" id="section2">
    <h2>Section Two</h2>
    <p>...</p>
  </div>
  <div role="tabpanel" class="tab-pane fade" id="section3">
    <h2>Section Three</h2>
    <p>...</p>
  </div>
</div>
```

There is one method on the Tab plugin: `$().tab('show')`. This must have either a `data-target` or an `href` attribute pointing to the `.tab-pane` container element in the DOM. There are four events, and they are called in this order:

▶ `hide.bs.tab`—Fires on the current active tab, and it fires when a new tab is to be shown. Use the `event.target` and `event.relatedTarget` to target the current active tab and the new, soon-to-be-active tab, respectively.

▶ `show.bs.tab`—This then fires on the to-be-opened tab. It fires when the tab is triggered to show but before it has been shown. Use the `event.target` and `event.relatedTarget` to target the current active tab and the previously active tab, respectively.

▶ `hidden.bs.tab`—This fires on the hidden tab, the previously active tab. It fires after a new tab is shown. Use the `event.target` and `event.relatedTarget` to target the previously active tab and the new active tab, respectively.

▶ `shown.bs.tab`—This fires on the newly active tab that was just shown. It fires after the tab has been shown. Use the `event.target` and `event.relatedTarget` to target the active tab and the previously active tab, respectively.

ScrollSpy

ScrollSpy sets up your page so the navigation highlights the exact area that the customer is reading. As she scrolls through the page, the highlight in the navigation changes. This plugin is especially useful in long pages with internal navigation, but it is also useful in modern designs where the entire site is contained on one page.

ScrollSpy works by "spying" on an element (usually `<body>`) that will be scrolled by the user. As that element scrolls, Bootstrap watches for IDs that correspond to references in the target Bootstrap navigation element. Then it adds the `.active` class to that element (and removes it from any previously active element) to highlight it.

Using ScrollSpy

The first thing you need to use the ScrollSpy plugin is a Bootstrap navigation bar. If you need a reminder of how to add a Bootstrap nav element, refer to Hour 12. That navigation needs to be visible on the page as you scroll, such as with the `.navbar-fixed-top` or `.navbar-fixed-bottom` classes. Make sure your navigation has a unique identifier.

The navigation must have resolvable ID targets that point to specific areas on the page. In other words, your navigation must include `<a href="#partOne">` links to somewhere defined on the page. ScrollSpy won't work with links to other web pages. Plus, the content needs to be visible on the page for ScrollSpy to work. If you have hidden content on the page that is not visible, ScrollSpy will ignore it and not highlight that navigation element.

Add the `data-spy="scroll"` attribute to the `<body>` tag, and add the `data-target="#myNav"` attribute pointing to your navigation element. Make sure your `<body>` tag has a relative position by adding `body { position: relative; }` to your style sheet.

CAUTION

Target the Nav Container, Not the Nav List

Best practices indicate that you should surround a navigation list with the `<nav>` tag and then place the `.nav` and `.navbar` styles on that. But some people like to place all those on the `<ul>` tag instead. However, this won't work with ScrollSpy. You need to have a container around your `<ul>` navigation list. This container is what you target with ScrollSpy.

You also can use JavaScript to use ScrollSpy. Listing 16.7 shows how to apply it to the `<body>` tag.

LISTING 16.7   ScrollSpy on the `<body>` Tag with JavaScript

```
$('body').scrollspy({ target: '#myNav' });
```

ScrollSpy has one method: `.scrollspy('refresh')`. Use it to refresh the page and update ScrollSpy. This is useful if you add or remove elements from the DOM and need to update the navigation dynamically.

ScrollSpy has one option: `offset`. This is the number of pixels to offset from the top when calculating the position of the scroll. The default value is 10, but you can use any number. Set this option with the `data-offset` attribute, or add it to the options array in your script.

There is only one event for ScrollSpy: `activate.bs.scrollspy`. It fires whenever ScrollSpy activates a new item. This enables you to add more actions than just highlighting the appropriate navigation element.

Using These Plugins Together

These plugins are well suited for use together. If you affix your navigation list to the page, you could then use ScrollSpy to highlight the area in which the customer is. You could affix a tabbed navigation to the top of your page and then use ScrollSpy within the tabs to highlight the relevant tab.

The most common combination is the one on the Bootstrap website: vertical sub-navigation is affixed to the side and then ScrollSpy highlights the areas. For Listing 16.8, I removed most of the placeholder text, but it shows the basic HTML and CSS to create a left navigation menu that bolds the sections as you enter them.

LISTING 16.8   ScrollSpy and Affix Together

```
<!DOCTYPE html>
<html lang="en">
  <head>
    <meta charset="utf-8">
    <meta http-equiv="X-UA-Compatible" content="IE=edge">
    <meta name="viewport"
          content="width=device-width, initial-scale=1">
    <title>Bootstrap Affix with ScrollSpy</title>

    <!-- Bootstrap -->
    <link href="css/bootstrap.min.css" rel="stylesheet">
    <!-- HTML5 shim and Respond.js for IE8 support of HTML5
    elements and media queries -->
    <!-- WARNING: Respond.js doesn't work if you view the page via
    file:// -->
    <!--[if lt IE 9]>
      <script
src="https://oss.maxcdn.com/html5shiv/3.7.2/html5shiv.min.js">
      </script>
      <script
src="https://oss.maxcdn.com/respond/1.4.2/respond.min.js"></script>
    <![endif]-->
    <style>
      body { position: relative; }
```

```
      .active { font-weight: bold; }
    </style>
  </head>
<body data-spy="scroll" data-target="#myNav">
<div class="container">
  <div class="col-md-3">
    <nav id="myNav">
      <ul class="nav" data-spy="affix">
        <li><a href="#section1">Section One</a></li>
        <li><a href="#section2">Section Two</a></li>
        <li><a href="#section3">Section Three</a></li>
      </ul>
    </nav>
  </div>
    <article class="col-md-9">
    <div id="section1">
      <h2>Section One</h2>
      <p>...</p>
    </div>
    <div id="section2">
      <h2>Section Two</h2>
      <p>...</p>
    </div>
    <div id="section3">
      <h2>Section Three</h2>
      <p>...</p>
    </div>
    </article>
</div>

<script src="http://code.jquery.com/jquery-latest.js"></script>
<script src="js/bootstrap.min.js"></script>

</body>
</html>
```

Don't hesitate to use any of the Bootstrap JavaScript plugins together on your page to create exactly the designs and interactivity you need.

Summary

This hour you learned about three Bootstrap plugins: Affix, Tab, and ScrollSpy. These plugins help you make more interesting and useful navigation elements as well as other features. You learned about how to fix elements on the screen while the rest of the content scrolls around it. You learned how to create hidden content with tabbed navigation, and you learned about how to highlight specific parts of your navigation when readers scroll to it with the ScrollSpy plugin.

Table 16.1 describes the options for these plugins. Table 16.2 explains the methods they use, and Table 16.3 describes the events available on them. These tables list only the options, methods, and events for the plugins that have them. Table 16.4 covers the CSS classes you learned about in this hour.

TABLE 16.1 Options

Plugin	Option	Value	Description
Affix	offset	A number, a function, or an object	Describes how many pixels to offset from the screen when calculating the position of the scroll. If you want to define different values for the top and bottom offset, provide an object—for example, offset { top: 5, bottom: 3 }. The default value is 10.
Affix	target	A selector, node, or jQuery element	Specifies the target element of the affix. The default value is the window object.
ScrollSpy	offset	A number	Pixels to offset from the top when calculating the position of the scroll. The default is 10.

TABLE 16.2 Methods

Plugin	Method	Description
ScrollSpy	.scrollspy('refresh')	Reloads ScrollSpy after an element has been added or removed from the DOM
Tab	.tab('show')	Activates a tab element and content container

TABLE 16.3 Events

Plugin	Event	Description
Affix	affix.bs.affix	This fires immediately before the element has been affixed.
Affix	affix-bottom.bs.affix	This fires immediately before the element hits the bottom of the affix.
Affix	affix-top.bs.affix	This fires immediately before the element hits the top of the affix.
Affix	affixed.bs.affix	This fires after the element is affixed.
Affix	affixed-bottom.bs.affix	This fires after the element hits the bottom affix.
Affix	affixed-top.bs.affix	This fires after the element hits the top affix.
ScrollSpy	activate.bs.scrollspy	This fires when the ScrollSpy activates.

Plugin	Event	Description
Tab	`hidden.bs.tab`	This fires after a new tab is shown and the previously active tab is hidden.
Tab	`hide.bs.tab`	This fires when a new tab is to be shown.
Tab	`show.bs.tab`	This fires on the tab shown but before the new tab has been shown.
Tab	`shown.bs.tab`	This fires on the tab shown after the new tab has been shown.

TABLE 16.4 CSS Classes

Class	Description
`.affix`	Adds the `position: fixed;` property to the element and defines it as an affixed element. Indicates that the element has been scrolled past.
`.affix-bottom`	Indicates that the affixed element has passed the bottom offset.
`.affix-top`	Indicates that the affixed element is in its topmost position.
`.in`	Indicates that the element should be displayed when fade is applied.
`.tab-content`	Indicates that the element contains tab content.
`.tab-pane`	Indicates that the element is a tab pane.

Workshop

The workshop contains quiz questions to help you process what you've learned in this hour. Try to answer all the questions before you read the answers.

Q&A

Q. Why do all the events have `.bs` in the name?

A. This is how Bootstrap tries to prevent collisions with other frameworks. By naming their events with both the framework name (`.bs`) and the plugin name (`.affix`), this ensures that other scripts that might have similar events don't use the same names.

Q. You always specify targets as IDs, but is it possible to identify target elements in other ways?

A. If you think of the target as a selector similar to jQuery selectors, you can use similar rules to select elements for your `data-target` values. For example, if you wanted to select an element with the class `.selectMe`, you would write `data-target=".selectMe"`. But remember that for some plugins the target needs to be unique on the page, and the Tab plugin changes the URL (if you don't prevent the default action). Thus, the URL needs to have a hash (#) target to work correctly.

Q. **When I used ScrollSpy, I didn't need a** `data-target`. **Is that attribute really required?**

A. Yes, it's required, but Bootstrap has a fallback option where it will apply ScrollSpy to the first navigation it finds if no target is specified. Even though this works, you run the risk of any changes breaking it.

Quiz

1. Which of these best describes what the Affix plugin does?

    a. Toggles the `position:fixed;` property on an element

    b. Forces an element to stay visible on the page

    c. Lets content scroll around an element

    d. Fixes content that would otherwise not display correctly

2. How do you affix an element?

    a. Add the attribute `data-toggle="affix"` to the element.

    b. Add the class `affix` to the element.

    c. Add the attribute `data-spy="affix"` to the element.

    d. Add the `affix.js` file to the page.

3. How can you set the top and bottom offsets on an Affixed element?

    a. Use the `data-offset-top` and `data-offset-bottom` attributes on the element.

    b. Use a JavaScript array `offset : { top: 60, bottom: 120 }` in the options.

    c. Use the `data-offset` attribute with two comma-separated values.

    d. Both A and B.

    e. None of the above.

4. True or False: The Tab plugin requires Bootstrap navigation.

5. Which of these is a Tab option?

    a. `offset`

    b. `target`

    c. `.tab('show')`

    d. None of the above

6. In the following code, what does the `e.preventDefault();` do?

```
$('#myTab a').click(function (e) {
  e.preventDefault();
  $(this).tab('show');
});
```

    a. Prevents the default action and blocks the tab from opening

    b. Prevents the default action and allows the tab to open

    c. Prevents the default action and does not place the hash in the URL

    d. Prevents the default action and places the hash in the URL

7. True or False: The ScrollSpy plugin requires Bootstrap navigation.

8. Which element is spied on with the ScrollSpy plugin?

    a. The `<html>` element

    b. The `<body>` element

    c. The `<nav>` element

    d. Whichever element needs to change

9. True or False: ScrollSpy can highlight based on URLs pointing anywhere.

10. When would you use the `.scrollspy('refresh')` method?

    a. Whenever the page refreshed.

    b. Whenever content changed on the page.

    c. Whenever DOM elements were removed or added.

    d. That is not a valid method for the ScrollSpy plugin.

Quiz Answers

1. a. Toggles the `position:fixed;` property on an element

2. c. Add the attribute `data-spy="affix"` to the element.

3. d. Both A and B.

4. False

5. d. None of the above. There are no options for the Tab plugin.

6. c. Prevents the default action and does not place the hash in the URL.

7. True

8. b. The `<body>` element

9. False

10. c. Whenever DOM elements were removed or added.

Exercises

1. Affix an object to your page. Play around with different elements, such as images or forms. What can you think of that would make a good floating element?

2. Create a tab pane for a page. Hide content behind three different panes, and then use the Tab plugin to show that content on click.

3. Add ScrollSpy to your navigation. If your navigation doesn't remain visible on the screen, test using a fixed navigation or use Affix to fix it in place.

Popovers and Tooltips

What You'll Learn in This Hour:

▶ How to add tooltips to Bootstrap pages
▶ The difference between tooltips and popovers
▶ How to add popovers
▶ The options, methods, and events for popovers and tooltips

Tooltips and popovers are small blocks of content that appear over and around the existing text. Tooltips typically show up on hover-over links or other text, while popovers appear on click and are usually larger.

In this hour you learn how to create tooltips and popovers on your websites to add information for your customers. You also learn how to add these features unobtrusively without annoying your readers.

Tooltips

A tooltip is a small text block that appears next to the triggering element. Bootstrap tooltips are written in small white text on a black, rounded-corner rectangle. They have a small triangle pointing to the element that triggered them. Figure 17.1 shows a sample tooltip.

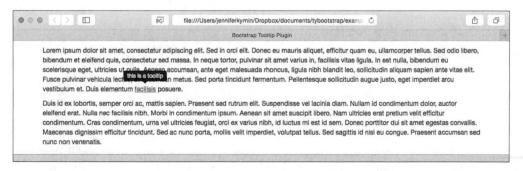

FIGURE 17.1
Standard tooltip.

Most tooltips are triggered when the mouse hovers over the element, but you also can trigger them with a click.

CAUTION

Tooltips Can Annoy Customers

You should bear in mind that many people find hovering tooltips annoying. However, tooltips that are triggered with a click can be confusing because most customers expect to be taken somewhere else when they click rather than just opening a small text box. Be sure to test your tooltips with your customers, especially if the tooltips convey important information.

To build a tooltip on a link, you must add `data-toggle="tooltip"` to your link (`<a>`) and ensure there is a `title` attribute with the text you want the tooltip to display. You can give your link an `href` value if you want it to go somewhere on click; otherwise, point to a hash tag: `href="#"`.

NOTE

Be Aware When Adding Tooltips to Hidden or Disabled Elements

You should not use a tooltip on elements that are hidden (`display: none;`) because the tooltip will be incorrectly positioned. Also, if you use a tooltip on a disabled element (with either the `disabled` attribute or the `.disabled` class), you should put a container `<div>` around the element and then put the tooltip on the container.

But before you rush out and add a tooltip to all your links, you need to do one more thing: initialize the tooltips.

The Tooltip and Popover plugins are opt-in for performance reasons. This is because they can be resource intensive. Plus, they often annoy customers as I mentioned in the previous Caution. So it's important that when you add them to pages you do so with some forethought. You are required to initialize tooltips in your scripts. Listing 17.1 shows JavaScript to select all tooltips with the `data-toggle` attribute.

LISTING 17.1   Initialize Tooltips

```
$(document).ready(function(){
  $(function () {
    $('[data-toggle="tooltip"]').tooltip();
  });
});
```

As with all other Bootstrap plugins, you need to include the Bootstrap JavaScript and jQuery. Listing 17.1 uses the document ready function jQuery function to ensure that everything is loaded before the tooltips are initialized. This will help your pages load more quickly.

When a tooltip is displayed on the page, it generates some HTML, as shown in Listing 17.2. Inside the HTML are several classes: .tooltip, .top, .tooltip-arrow, and .tooltip-inner. You can use these classes in your CSS to provide more styles to the tooltips.

LISTING 17.2 Tooltip-Generated HTML

```
<div class="tooltip top" role="tooltip">
  <div class="tooltip-arrow"></div>
  <div class="tooltip-inner">
    Some tooltip text!
  </div>
</div>
```

The one class that changes is top because this defines where the tooltip will display. The default is top, but with the placement option, you can position your tooltip on the top, right, bottom, or left. You also can use the keyword auto on your placement to tell the browser to position the tooltip dynamically. For example, if the placement is "auto top", the tooltip will display on the top if possible; otherwise, it will display on the bottom. This is particularly useful if the triggering element is near the edges of the screen.

TRY IT YOURSELF ▼

Add a Tooltip on a Button

Tooltips can be added to any element that can be interacted with on the page. This is most commonly the <a> tag and button elements. It's a good idea to stick with those for your tooltips so your customers understand that they are interactive. In this Try It Yourself you learn how to add a tooltip to a button that appears when the button is clicked.

1. Open your bootstrap web page or create a new page with a bootstrap template.

2. Add a button somewhere on the page:

   ```
   <button type="button" class="btn btn-default">
     Click if you love Dandelions
   </button>
   ```

3. Add a title attribute to the button, and include the text you want your tooltip to display.

4. Add the data-toggle="tooltip" attribute to the button.

5. Add data-trigger="click" to the button to change the trigger to when the button is clicked.

6. Add any other options as data-* attributes, such as the data-placement attribute to change the position of the tooltip.

7. Test your tooltip in a browser. It won't work, and the title will show up when you mouse over the button.

8. Add a `<script>` at the bottom of the document to initialize the tooltip:

```
<script>
$(document).ready(function(){
  $(function () {
    $('[data-toggle="tooltip"]').tooltip();
  });
});
</script>
```

9. Test it again. This time it will work.

Listing 17.3 shows the HTML for Figure 17.2 showing a tooltip when a button is clicked.

LISTING 17.3 Tooltip HTML

```
<!DOCTYPE html>
<html lang="en">
  <head>
    <meta charset="utf-8">
    <meta http-equiv="X-UA-Compatible" content="IE=edge">
    <meta name="viewport"
          content="width=device-width, initial-scale=1">
    <title>Bootstrap Tooltip on a Button</title>

    <!-- Bootstrap -->
    <link href="css/bootstrap.min.css" rel="stylesheet">
    <!-- HTML5 shim and Respond.js for IE8 support of HTML5
    elements and media queries -->
    <!-- WARNING: Respond.js doesn't work if you view the page via
    file:// -->
    <!--[if lt IE 9]>
      <script
src="https://oss.maxcdn.com/html5shiv/3.7.2/html5shiv.min.js">
      </script>
      <script
src="https://oss.maxcdn.com/respond/1.4.2/respond.min.js"></script>
    <![endif]-->
    <style>
    body { padding-top: 70px; }
    img#dandylionLogo { height:100%; width: auto; display: inline;
                        margin-top: -10px; }
    </style>

  </head>
```

```
<body>
    <nav class="navbar navbar-default navbar-fixed-top">
      <div class="container-fluid">
        <div class="navbar-header">
          <button type="button" class="navbar-toggle collapsed"
                  data-toggle="collapse" data-target="#collapsedNav">
            <span class="sr-only">Toggle navigation</span>
            <span class="icon-bar"></span>
            <span class="icon-bar"></span>
            <span class="icon-bar"></span>
          </button>
          <a href="#" class="navbar-brand">
            <img src="images/dandylion-logo.png" alt="Dandylion"
                id="dandylionLogo" />Dandylions</a>
        </div>

        <div class="collapse navbar-collapse" id="collapsedNav">
          <ul class="nav navbar-nav">
            <li class="active"><a href="#">Products</a></li>
            <li><a href="#">Services</a></li>
            <li><a href="#">Support</a></li>
            <li class="dropdown">
            <a href="#" class="dropdown-toggle"
              data-toggle="dropdown" role="button"
              aria-expanded="false">
            About <span class="caret"></span>
            </a>
            <ul class="dropdown-menu" role="menu">
              <li><a href="#">Articles</a></li>
              <li><a href="#">Related Sites</a></li>
            </ul>
            </li>
          </ul>
          <p class="navbar-text">We love
          <a href="" class="navbar-link">Weeds</a>!</p>
          <form class="navbar-form navbar-right" role="search">
            <div class="form-group">
              <input type="text" class="form-control"
                    placeholder="Search">
            </div>
          </form>
          <button type="button"
                  class="btn btn-default navbar-btn navbar-right">
            Contact Us
          </button>
        </div>
```

```
      </div>
   </nav>
   <div class="container">
     <div class="row">
     <article id="mainarticle" class="col-lg-5 col-md-6 col-xs-12">
       <h3>What is a Dandelion?</h3>
       <p><span class="pronounce"><strong><em>Taraxacum</em></strong>
       <em>/təˈræksəkʉm/</em></span> is a large genus of
       flowering plants in the family Asteraceae. They are native
       to Eurasia and North and South America, and two species,
       T. officinale and T. erythrospermum, are found as weeds
       worldwide. Both species are edible in their entirety.
       The common name
       <span class="pronounce"><strong>dandelion</strong>
       <em>/ˈdændɨlaɪ.ən/</em></span> (dan-di-ly-ən, from French
       dent-de-lion, meaning "lion's tooth") is given to members
       of the genus, and like other members of the Asteraceae
       family, they have very small flowers collected together
       into a composite flower head. Each single flower in a head
       is called a floret. Many <em>Taraxacum</em> species produce
       seeds asexually by apomixis, where the seeds are
       produced without pollination, resulting in offspring that
       are genetically identical to the parent plant.</p>
       <button type="button" class="btn btn-default"
               data-toggle="tooltip" data-placement="top"
               data-trigger="click"
               title="Everyone should love Dandelions!">
         Click if you love Dandelions
       </button>
     </article>
     <aside id="links" class="col-md-3 col-xs-12">
       <h3>Learn More About Dandelions</h3>
       <ul>
         <li><a href="#">Do Dandelions Burn in Colors?</a></li>
         <li><a href="#">Top 5 Things To Do With
         Dandelion</a></li>
         <li><a href="#">How to Grow and Harvest Dandelion
         Greens</a></li>
         <li><a href="#">Dandelion Honey Recipe</a></li>
         <li><a href="#">Get Rid of Dandelions the Smart
         Way@#8212Eat Them!</a></li>
       </ul>
     </aside>
     <aside id="sidebar" class="col-lg-3 col-md-3 col-xs-12">
       <p><img src="images/dandy.jpg" class="img-responsive"
               alt=""/></p>
```

```
      <p><img src="images/seeded.jpg" class="img-responsive"
            alt=""/></p>
    </aside>
    <footer class="col-xs-12">
      <p>Text from
      <a href="http://en.wikipedia.org/wiki/Dandelion">
        Wikipedia
      </a></p>
    </footer>

    </div>
  </div>
<script src="http://code.jquery.com/jquery-latest.js"></script>
<script src="js/bootstrap.min.js"></script>
<script>
$(document).ready(function(){
  $(function () {
    $('[data-toggle="tooltip"]').tooltip();
  });
});
</script>
</body>
</html>
```

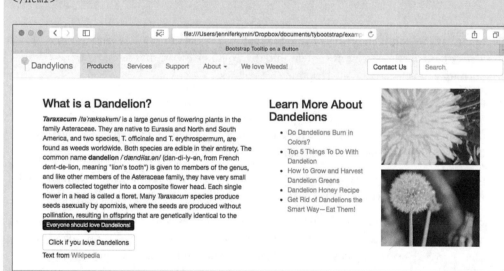

FIGURE 17.2
A tooltip on a button.

CAUTION

Button and Input Group Tooltips Require the container **Option**

If you are going to put tooltips on elements inside a .btn-group or .input-group, you have to specify the option container: 'body'. If you don't, you might find the element grows wider or loses its rounded corners when the tooltip is triggered.

Tooltip Options

The Tooltip plugin offers several options you can use to adjust how your tooltips work:

▶ animation—This applies a CSS fade transition to the tooltip. The value is true by default, and false turns off the fade.

▶ container—This is false by default, but if turned on it allows you to append the tooltip to a specific element (for example, container: 'body'). This allows you to position the tooltip in the flow of the document near the triggering element. This prevents the tooltip from moving away from the triggering element when the window is resized.

▶ delay—Defines a delay in milliseconds for how long until the tooltip is shown and hidden. If you write it as an object, you can define different values for showing and hiding (for example, delay: { "show": 200, "hide": 400 }). This will not apply when the trigger is manual.

▶ html—Allows you to insert HTML into the tooltip. The default is false and uses the jQuery text method to insert content into the DOM. Be aware that setting it to true can open up your site to cross-site scripting (XSS) attacks.

▶ placement—This indicates how to position the tooltip. The values are top, right, bottom, left, and auto. You also can use a function to determine the placement. This calls the function with the tooltip DOM node as its first argument and the triggering element DOM node as its second. The this context is set to the tooltip instance.

▶ selector—When a selector is provided, this allows tooltips to be placed on elements that were dynamically created in the DOM using the jQuery on function. The default is false.

▶ template—If you don't like the base HTML for the tooltips, you can change it with the template option. The tooltip's title will be injected into the .tooltip-inner element. The .tooltip-arrow becomes the arrow, and it should all be wrapped by an element with the .tooltip class. Listing 17.2 shows the default HTML.

▶ title—If the title attribute is missing, this is the default value for the tooltip. The default is an empty string: ''. You also can set it to a function where the this reference is set to the element to which the tooltip is attached.

- ► trigger—This defines how the tooltip is triggered. Possible values are `click`, `hover`, `focus`, and `manual`. You can pass multiple triggers separated by spaces. The default value is `hover focus`.

- ► viewport—This keeps the tooltip within the bounds of this element. The default is `{ selector: 'body', padding: 0 }`.

You can define all the options using `data-*` attributes or directly in your JavaScript.

Tooltip Methods

The Tooltip plugin provides five methods you can use to work with the tooltips:

- ► `$().tooltip(options)`—This attaches a tooltip handler to an element collection with the options you define.

- ► `.tooltip('show')`—This reveals the tooltip and returns to the caller before the tooltip has been shown. This is considered a manual triggering of the tooltip.

- ► `.tooltip('hide')`—This hides the element's tooltip. It returns to the caller before the tooltip has been hidden. This is considered a manual triggering of the tooltip.

- ► `.tooltip('toggle')`—This toggles the element's tooltip on or off. It returns to the caller before the tooltip has been shown or hidden. This is considered a manual triggering of the tooltip.

- ► `.tooltip('destroy')`—This hides and destroys the element's tooltip. Tooltips that are created using the `selector` option cannot be individually destroyed on descendant trigger elements.

Tooltip Events

Four events occur when the Tooltip plugin is used:

- ► `show.bs.tooltip`—This fires immediately when the `show` method is called.

- ► `shown.bs.tooltip`—This fires when the tooltip has been made visible to the user. It waits until any CSS transitions are complete.

- ► `hide.bs.tooltip`—This event fires immediately when the `hide` method is called.

- ► `hidden.bs.tooltip`—This event fires when the tooltip is completely hidden from the user, including any CSS transitions.

Popovers

In many ways, popovers are just like tooltips. They add overlays of content onto the web page when they are triggered. In fact, to use popovers, you have to have the Tooltip plugin enabled as well. But popovers give you a few more formatting options that aren't available with tooltips. Figure 17.3 shows a basic popover.

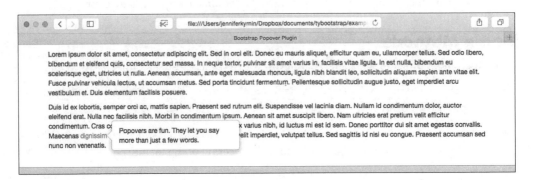

FIGURE 17.3
A basic popover.

To add a popover to your page, you need to include both the `popover.js` and the `tooltip.js` scripts, or you can use the bootstrap.js file. Similar to tooltips, you must initialize the popovers for them to work. The easiest way to initialize them is to select them by the `data-toggle` attribute just like you did with tooltips. Listing 17.4 shows an initialization script.

LISTING 17.4   Initialize Popovers

```
<script>
$(document).ready(function(){
  $(function () {
    $('[data-toggle="popover"]').popover();
  });
});
</script>
```

After you've initialized the popovers, you can add them to your HTML. They are usually added to buttons. The biggest difference between popovers and tooltips is that popovers appear on click and remain on screen until the customer clicks the trigger again. Buttons make this functionality more obvious. Listing 17.5 shows the HTML to add a default popover.

LISTING 17.5   Popover on a Button

```
<button type="button" class="btn btn-default" data-toggle="popover"
      title="Popovers are fun"
      data-content="You can add titles and they stick.">
  Click to see a Popover
</button>
```

You need just two attributes to create a simple popover: `data-toggle` and `data-content`. The `title` attribute adds a title to your popover, as shown in Figure 17.4.

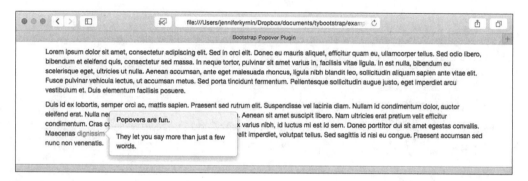

FIGURE 17.4
A basic popover with a title.

If you want the popover to disappear with the next click—whether it's on the trigger button or not—you can do that with the `trigger` option. Change the trigger to `focus`, but there are a few more things you need to do to get the dismiss-on-next-click popover to work correctly on all browsers:

▶ You must use the `<a>` tag and not the `<button>` tag.

▶ You need to include the `role="button"` attribute.

▶ You should include the `tabindex` attribute, but it can be set to whatever value you like.

LISTING 17.6   Dismiss-On-Next-Click Popover

```
<a href="#" class="btn btn-default" role="button"
   data-toggle="popover" data-trigger="focus"
   title="Make this Popover Disappear"
   data-content="Click anywhere and the popover will leave."
   tabindex="0">
  This is a Dismissable Popover
</a>
```

Add a Popover to a Page

It's easy to add a popover. This Try It Yourself takes you through the steps to add a popover to a button on a web page:

1. Open a Bootstrap web page in your web editor.

2. Add a button to trigger your popover:

   ```
   <button role="button" class="btn btn-default">
     Dandelion Fun Fact
   </button>
   ```

3. Add the `data-toggle="popover"` attribute to trigger a popover.

4. Add a title with the `title` attribute.

5. Add content in the `data-content` attribute.

6. Add styles to style the `.popover`, `.popover-title`, and `.popover-content` classes.

Figure 17.5 shows my popover in action.

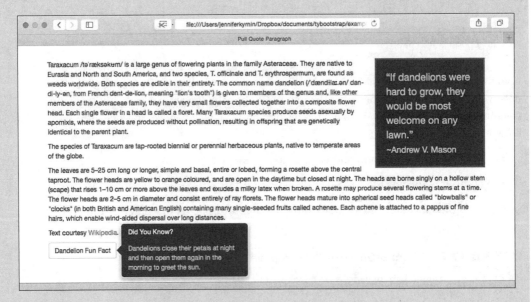

FIGURE 17.5
My styled popover.

When a popover is triggered, the HTML in Listing 17.7 is added to the DOM. This includes the classes .popover, .popover-title, and .popover-content as well as the .arrow class. These give the popover its rounded corners, drop shadow, and background colors. You also can use them to style the popover to match your site design.

LISTING 17.7   The Default Popover Template

```
<div class="popover" role="tooltip">
  <div class="arrow"></div>
  <h3 class="popover-title"></h3>
  <div class="popover-content"></div>
</div>
```

Popover Options

A lot of options are available that you can use with either the data-* attributes or directly in the JavaScript to adjust how your popovers work:

▶ animation—Adds a CSS fade transition to the popover. The default is true.

▶ container—This is false by default, but if turned on it allows you to append the pop-over to a specific element (for example, container: 'body'). This enables you to position the popover in the flow of the document near the triggering element. This prevents the popover from moving away from the triggering element when the window is resized.

▶ content—Provides default content if the data-content attribute is not present. It can be built with a JavaScript function, and it will be called with the this reference set to the element to which the popover is attached.

▶ delay—The number of milliseconds to delay showing and hiding the popover. You also can use an object such as delay: { "show": 300, "hide": 1000 }. The default is 0.

▶ html—Enables you to insert HTML into the popover. The default is false and uses the jQuery text method to insert content into the DOM. Be aware that setting it to true can open up your site to cross-site scripting (XSS) attacks.

▶ placement—Positions the popover to the top, right, bottom, or left. You can use the term auto, which automatically reorients the popover when required. For example, data-placement="auto right" places the popover on the right unless there isn't enough room; then it places it on the left. The default value is right.

▶ selector—When a selector is provided, this allows popovers to be placed on elements that were dynamically created in the DOM using the jQuery on function. The default is false.

▶ template—If you don't like the base HTML for the popovers, you can change it with the template option. The popover's title will be injected into the .popover-title element.

The popover's content will be injected into the `.popover-content` element. The `.arrow` element becomes the arrow, and it should all be wrapped by an element with the `.popover` class. Listing 17.7 shows the default HTML.

▶ `title`—Provides a default title if the `title` attribute is not present. It can be built with a JavaScript function and will be called with its `this` reference set to the element to which the popover is attached.

▶ `trigger`—Defines how the popover is triggered. The possible values are `click`, `hover`, `focus`, and `manual`. You can pass several triggers separated by a space to allow your popover to be triggered in multiple ways. The default value is `click`.

▶ `viewport`—This keeps the popover within the bounds of this element. The default is `{ selector: 'body', padding: 0 }`.

Popover Methods

The Popover plugin provides five methods you can use to work with popovers:

▶ `$().popover(options)`—This initializes the popovers for an element collection.

▶ `.popover('show')`—This reveals the popover. It returns to the caller before the popover has been shown, which is considered a manual triggering of the popover. Popovers with both empty titles and content are never displayed.

▶ `.popover('hide')`—This hides the element's popover. It returns to the caller before the popover has been hidden. This is considered a manual triggering of the popover.

▶ `.popover('toggle')`—This toggles the element's popover on or off. It returns to the caller before the popover has been shown or hidden, which is considered a manual triggering of the popover.

▶ `.popover('destroy')`—This hides and destroys the element's popover. Popovers that are created using the `selector` option cannot be individually destroyed on descendant trigger elements.

Popover Events

Four events occur when the Popover plugin is used:

▶ `show.bs.popover`—This fires immediately when the `show` method is called.

▶ `shown.bs.popover`—This fires when the popover has been made visible to the user. It waits until any CSS transitions are complete.

▶ `hide.bs.popover`—This event fires immediately when the `hide` method is called.

▶ `hidden.bs.popover`—This event fires when the popover is completely hidden from the user, including any CSS transitions.

Summary

This hour you learned about two plugins that give your pages a dynamic flare and provide additional information to readers: Tooltip and Popover. These plugins create small text boxes that pop up and provide additional information when the user triggers them.

Table 17.1 describes the options for these plugins. Table 17.2 explains the methods they use, and Table 17.3 describes the events available on them. Table 17.4 covers the CSS classes you learned about in this hour.

TABLE 17.1 Options

Plugin	Option	Value	Description
Popover, Tooltip	animation	true or false	Applies a CSS fade to the popover or tooltip. The default is true.
Popover, Tooltip	container	A string or false	Appends the popover or tooltip to a specific element. The default is false.
Popover	content	A string or a function	Default content to be used if the data-content attribute is not present. The default is an empty string.
Popover, Tooltip	delay	A number or an object	The number of milliseconds to delay showing and hiding the popover or tooltip. The default is 0.
Popover, Tooltip	html	true or false	Allows you to insert HTML into the popover or tooltip. The default is false.
Popover, Tooltip	placement	top, right, bottom, left, and auto	Defines where the popover or tooltip should display. The default is right.
Popover, Tooltip	selector	A string or false	Enables dynamic HTML elements to have popover and tooltips added to them. The default is false.
Popover, Tooltip	template	See Listings 17.2 and 17.7	The base HTML to create a popover or tooltip.
Popover, Tooltip	title	A string or function	Value of the title if one is not specified with the title attribute.
Popover, Tooltip	trigger	click, hover, focus, or manual	How a popover or tooltip is triggered. The default is click for popovers and hover focus for tooltips.
Popover, Tooltip	viewport	A string or an object	Keeps the tooltip or popover within the bounds of this element.

TABLE 17.2   Methods

Plugin	Method	Description
Popover	`$().popover(options)`	Initializes popovers
Popover	`.popover('destroy')`	Removes and destroys the popover
Popover	`.popover('hide')`	Hides the popover
Popover	`.popover('show')`	Shows the popover
Popover	`.popover('toggle')`	Changes the visibility of the popover to show or hide
Tooltip	`$().tooltip(options)`	Initializes tooltips
Tooltip	`.tooltip('destroy')`	Removes and destroys the tooltip
Tooltip	`.tooltip('hide')`	Hides the tooltip
Tooltip	`.tooltip('show')`	Shows the tooltip
Tooltip	`.tooltip('toggle')`	Changes the visibility of the tooltip to show or hide

TABLE 17.3   Events

Plugin	Event	Description
Popover	`hide.bs.popover`	This fires immediately as the `hide` method is called.
Popover	`hidden.bs.popover`	This fires when the popover is finished being hidden.
Popover	`show.bs.popover`	This fires immediately as the `show` method is called.
Popover	`shown.bs.popover`	This fires when the popover is finished being shown.
Tooltip	`hide.bs.tooltip`	This fires immediately as the `hide` method is called.
Tooltip	`hidden.bs.tooltip`	This fires when the tooltip is finished being hidden.
Tooltip	`show.bs.tooltip`	This fires immediately as the `show` method is called.
Tooltip	`shown.bs.tooltip`	This fires when the tooltip is finished being shown.

TABLE 17.4   CSS Classes

Class	Description
`.arrow`	Adds an arrow to a popover pointing at the element to which it's attached. This is generated by the plugin.
`.bottom`	Positions the tooltip on the bottom. This is generated by the plugin.
`.left`	Positions the tooltip on the left. This is generated by the plugin.
`.popover`	Indicates this is a popover container. This is generated by the plugin.
`.popover-content`	Indicates this is popover content. This is generated by the plugin.
`.popover-title`	Indicates this is a popover title. This is generated by the plugin.

Class	Description
`.right`	Positions the tooltip on the right. This is generated by the plugin.
`.tooltip`	Indicates this is a tooltip container. This is generated by the plugin.
`.tooltip-arrow`	Adds an arrow to the tooltip pointing at the element to which it's attached. This is generated by the plugin.
`.tooltip-inner`	Indicates this is the tooltip contents. This is generated by the plugin.
`.top`	Positions the tooltip to the top. This is generated by the plugin.

Workshop

The workshop contains quiz questions to help you process what you've learned in this hour. Try to answer all the questions before you read the answers.

Q&A

Q. What is the difference between a modal, a popover, and a tooltip?

A. There are many ways to present information on websites, and these are all methods of getting a reader's attention and presenting new information. The main difference between a modal and a tooltip or popover is that modals want the user to take some action. They often ask a question or present a form to be filled in. Popovers and tooltips just present some information. You can learn more about modals in Hour 15, "Modal Windows."

Q. Popovers and tooltips seem very similar, but are they really different things?

A. In the Bootstrap world, there are two main differences between a popover and a tooltip. A popover is triggered by a click and has a headline as well as main text. A tooltip is triggered by hover or focus and just has text. In the real world, Bootstrap popovers are just fancy tooltips.

Q. Is it a good idea to use tooltips or popovers? Don't people find them annoying?

A. This will depend on your website. But in general, the more tooltips you have, the more annoying they will be to users. And the larger the tooltips or popovers are, the more content they will cover, which will make them more annoying.

Be aware of what users expect. If they click a link, they expect to be taken to another location. If all that happens is a tooltip appears, they could get annoyed. Use tooltips and popovers where it adds value to your pages, but don't go overboard. If you can, do user testing to discover whether your customers find them annoying.

Quiz

1. Which of the following triggers is the standard tooltip trigger?

    a. Clicking on an element

    b. Mousing over an element

    c. Mousing near an element

    d. Clicking near an element

2. True or False: You can only put tooltips on links (`<a>` tags).

3. Why is it a bad idea to use tooltips on elements with the `display: none;` style?

    a. The style can be transferred to the tooltip, hiding it, too.

    b. It won't work unless you put it on a container around the hidden element.

    c. They will end up displaying in the wrong place.

    d. It's not a bad idea. That works fine.

4. What do you need to add besides the `data-toggle="tooltip"` attribute to get a tooltip to display?

    a. Nothing, `data-toggle="tooltip"` is all you need.

    b. The `title` attribute, with the text for the tip.

    c. The `title` attribute, and you need to initialize it.

    d. Nothing; `data-toggle="tooltip"` is not the attribute to turn on tooltips.

5. What does the `.auto` class do to a tooltip?

    a. It makes the tooltips appear automatically.

    b. It automates the generation of tooltips.

    c. It changes the color of the tooltip.

    d. It changes the position of the tooltip, if necessary.

6. What is the difference between a tooltip and a popover?

    a. A popover adds more formatting options.

    b. A tooltip adds more formatting options.

    c. A popover adds more positioning options.

    d. A tooltip adds more positioning options.

7. Which scripts are required to use popovers?

    a. The `popover.js` file

    b. The `tooltip.js` file

    c. A link to jQuery

    d. All of the above

8. Which of these holds the popover content?

    a. `.popover`

    b. `.popover-inner`

    c. `.popover-content`

    d. `.popover-title`

9. Which event fires as a popover is being shown?

    a. `hide.bs.popover`

    b. `show.bs.popover`

    c. `shown.bs.popover`

    d. None

10. What can you do if you want to use an `<aside>` to contain your popovers rather than a `<div>`?

    a. Change the popover with the `data-template` attribute.

    b. Modify the popover with the `data-html` attribute.

    c. Use the `popover.html` method.

    d. Nothing. You can't change that.

Quiz Answers

1. b. Mousing over an element.

2. False. You can also put them on `<button>` tags.

3. c. They will end up displaying in the wrong place.

4. c. The `title` attribute, and you need to initialize it.

5. d. It changes the position of the tooltip, if necessary.

6. a. A popover adds more formatting options.

7. d. All of the above.

8. c. `.popover-content`

9. b. `show.bs.popover`

10. a. Change the popover with the `data-template` attribute.

Exercises

1. Add a tooltip to a link on your page. Make sure that it adds value to the link and isn't too long. Use the tooltip CSS classes (refer to Table 17.4) to style the tooltip to fit your website design.

2. Add a popover that disappears on the next click.

HOUR 18
Transitions, Buttons, Alerts, and Progress Bars

What You'll Learn in This Hour:

▶ What Bootstrap transitions are for

▶ How to adjust your buttons

▶ Adding alerts and making them dismissible

▶ Creating progress bars

Bootstrap offers a lot of features you see on dynamic pages. The plugins covered this hour are ones you often see on pages that change a lot or provide information over time to the customers.

This hour covers the `transition.js`, `button.js`, and `alert.js` plugins as well as the progress bars component. These plugins and components help you make things more dynamic for your customers.

Transitions

Transitions are the animation effects that take place when something changes on your page. Several of the earlier hours mentioned transitions such as fading in and fading out.

Most transitions are done with CSS, not JavaScript. This plugin is in Bootstrap to help improve how transitions work when the CSS transitions don't, such as in older browsers. It is based on the Modernizr (http://modernizr.com/) transitions support. It works as a helper for the `transitionEnd` event and emulates CSS transitions in browsers that don't support them.

Transitions are used for things like the following:

▶ Sliding in modals (refer to Hour 15, "Modal Windows")

▶ Fading tabs in and out (refer to Hour 16, "Affix, Tab, and ScrollSpy")

▶ Animating tooltips and popovers (refer to Hour 17, "Popovers and Tooltips")

▶ Fading alerts in and out (see the section "Alerts" later in this hour)

▶ Sliding in collapsed content (see Hour 19, "Collapse and Accordion")

▶ Sliding between carousel panes (see Hour 20, "Carousels")

The nice thing about the `transition.js` plugin is that you don't have to explicitly call it when using other Bootstrap plugins. But if you want to have nice-looking transitions in all cases, be sure to include `transition.js` in your compile or use the full `bootstrap.js` file.

Buttons

Buttons, as you learned in Hour 11, "Styling and Using Buttons and Button Groups," are a powerful Bootstrap feature. But you can do more than simply place buttons on your page. With the `button.js` plugin, you can control the state of the button, toggle the button on or off, and even turn radio buttons and checkboxes into buttons. To use it, you just need the `bootstrap.js` file and jQuery.

Button States

If something is going to take some time after a click, it can be useful to provide some type of indicator to your customers. You can change the state of the button to indicate that something is loading or provide other text strings for other states. Figure 18.1 shows a button that has been changed to a loading state.

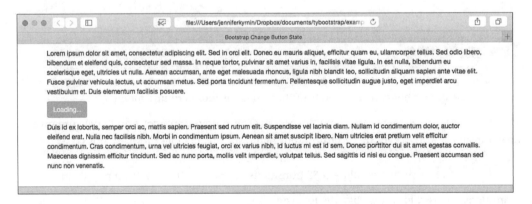

FIGURE 18.1
A button with a changed state.

Changing the state of your button uses the `$().button(string)` method. This swaps the text of the button to another `data-*` defined text state. For example, the button in Figure 18.1 uses a `data-loading-text` attribute. The button is written as in Listing 18.1, and the script is written as in Listing 18.2.

LISTING 18.1   A Button That Changes State

```
<button type="button" id="myButton" data-loading-text="Loading..."
        class="btn btn-primary" autocomplete="off">
  Loading state
</button>
```

LISTING 18.2   The Script to Change Button State

```
<script>
  $('#myButton').on('click', function () {
    var $btn = $(this);
    $btn.button('loading');
  })
</script>
```

You can set multiple states on your buttons. In the data-* attributes, instead of loading, you would write another word, such as complete (for example, data-complete-text="Done! Phew!"). Then you can write your script so the button changes text as different things happen.

CAUTION

Firefox Saves State Across Page Loads

Firefox saves the state of forms even if the page is reloaded. This is done to assist readers so they don't lose their form data, but if your button states need to reset when the page reloads, this can cause problems. One workaround is to use the autocomplete="off" attribute. This is documented in Mozilla bug #654072 (https://bugzilla.mozilla.org/show_bug.cgi?id=654072).

Toggle Button

You can activate toggling on a single button with the data-toggle="button" attribute. This sets the button to "clicked" and then "unclicked" as you turn it on and off. Listing 18.3 shows the HTML for a basic toggle button.

LISTING 18.3   A Standard Toggle Button

```
<button type="button" id="myButton" data-toggle="button"
        class="btn btn-primary" aria-pressed="false"
        autocomplete="off">
  Loading state
</button>
```

You also can set your button to be pre-toggled by adding the .active class and the aria-pressed="true" attribute, as in Listing 18.4.

LISTING 18.4 A Pre-Toggled Button

```
<button type="button" id="myButton" data-toggle="button"
        class="btn btn-primary active" aria-pressed="true"
        autocomplete="off">
  Loading state
</button>
```

Checkbox and Radio Buttons

If you have a button group (refer to Hour 11) that contains checkbox or radio button input controls, you can turn them into selectable buttons with the button.js plugin.

Place the attribute data-toggle="buttons" on the button group container <div>. Then surround the <input> tags with a <label> and define them as buttons with the btn classes.
Listing 18.5 shows how the HTML for checkbox buttons will look, and Figure 18.2 shows a styled
version of that HTML on a web page.

LISTING 18.5 Checkboxes as Buttons

```
<div class="btn-group" data-toggle="buttons">
  <label class="btn btn-primary active">
    <input type="checkbox" autocomplete="off" checked>Dandelions!
  </label>
  <label class="btn btn-primary">
    <input type="checkbox" autocomplete="off">Roses
  </label>
  <label class="btn btn-primary">
    <input type="checkbox" autocomplete="off">Orchids
  </label>
</div>
```

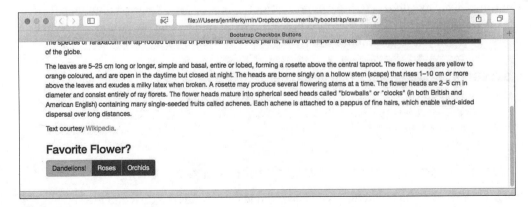

FIGURE 18.2
Checkboxes as buttons.

Radio buttons work in the same way, but only one button toggles on at a time. Listing 18.6 shows a radio button group.

LISTING 18.6   Radio Input Types as Buttons

```
<div class="btn-group" data-toggle="buttons">
  <label class="btn btn-primary active">
    <input type="radio" name="options" id="option1"
           autocomplete="off" checked>Dandelions!
  </label>
  <label class="btn btn-primary">
    <input type="radio" name="options" id="option2"
           autocomplete="off">Roses
  </label>
  <label class="btn btn-primary">
    <input type="radio" name="options" id="option3"
           autocomplete="off">Orchids
  </label>
</div>
```

Button Methods

The three methods in the `button.js` plugin are

▶ `$().button('toggle')`—Toggles the state of the button to make it appear activated (or deactivated)

▶ `$().button('reset')`—Resets the button state changing the text back to the original text

▶ `$().button(string)`—Swaps the button text to the `data-*` attribute defined by the string (for example, `$().button('loading')` points to the attribute `data-loading-text="Loading...")`

Alerts

Alerts are messages that let the user know something happened. They appear on the screen and typically include a dismiss button in the upper-right corner. Alerts can contain any HTML you need in them, but best practices recommend that you keep them as simple as possible.

Alerts are created using the `.alert` class and a contextual class. The four contextual classes for alerts are

▶ `.alert-success`

▶ `.alert-info`

▶ `.alert-warning`

▶ `.alert-danger`

There is no default class for alerts, so you must include a contextual class. Also, as with all contextual classes, be sure to include visible text inside the alert to keep it accessible. Listing 18.7 shows the HTML for four different alerts. These are pictured in Figure 18.3.

LISTING 18.7   Four Different Alerts

```
<div class="alert alert-success" role="alert">
  Success!
</div>
<div class="alert alert-info" role="alert">
  Info
</div>
<div class="alert alert-warning" role="alert">
  Warning
</div>
<div class="alert alert-danger" role="alert">
  Danger!
</div>
```

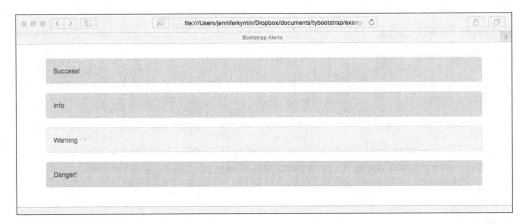

FIGURE 18.3
Four types of alerts.

If you want your customers to be able to dismiss the alert after they have read it, add the `.alert-dismissible` class to the alert and add a close button inside the alert. On the close button, add the `data-dismiss="alert"` attribute. Listing 18.8 shows a dismissible alert.

NOTE

Alternative Spelling

The standard spelling of dismissible by Bootstrap is with an *i* as in `.alert-dismissible`. However, there is also an alternative spelling in the CSS that has the same effect; it's spelled with an *a*, as in `.alert-dismissable`.

LISTING 18.8 Dismissible Alert

```
<div class="alert alert-warning alert-dismissible" role="alert">
  <button type="button" class="close" data-dismiss="alert"
          aria-label="Close">
    <span aria-hidden="true">&times;</span>
  </button>
  Warning! Learning about dandelions can be addictive.
</div>
```

To make the alert dismissible, you need to include the `alert.js` file or the full `bootstrap.js` file. You also need jQuery just like with the other Bootstrap plugins.

NOTE

Use a `<button>` Tag for the Close Button

Although you can use other tags to create a close button, you should use the `<button>` tag to ensure it works across all browsers. Also be sure to include the `aria-label="close"` attribute to keep your button accessible.

If you have links inside your alerts, you should use the `.alert-link` class on the link. This will provide matching colored links to the alert contextual class. Figure 18.4 shows some links inside alerts.

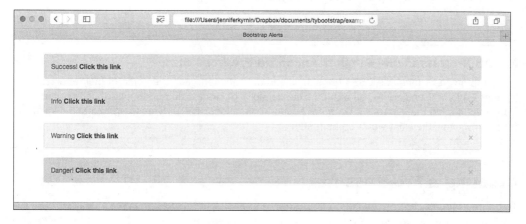

FIGURE 18.4
Four types of alerts with links.

If you want to add a transition animation, you can add the `.fade` and `.in` classes to the `.alert` element.

Alert Methods

The two methods you can use with the `alert.js` plugin are

▶ `$().alert()`—Makes an alert listen for click events on descendant elements that have the `data-dismiss="alert"` attribute. It wraps the selected alert with the close functionality. But it isn't necessary when using the data-api's autoinitialization.

▶ `$().alert('close')`—Closes the alert by removing it from the DOM. If a fade transition (with the `.fade` and `.in` classes) is set, it will fade out before being removed.

Alert Events

In addition, two events are available that you can use to hook into the alert functionality:

▶ `close.bs.alert`—This event fires immediately as the `close` method is called.

▶ `closed.bs.alert`—This event fires after the alert is closed. It waits until after any CSS transitions are complete.

Progress Bars

Progress bars are a useful component of Bootstrap that provide information for how far along a process is until it completes. You can use static progress bars on things like multipage forms or long articles to give readers a clue as to how far along they are. You also can set up dynamic progress bars that update periodically on a dynamic page.

CAUTION

Not All Browsers Fully Support Animated Progress Bars

Bootstrap progress bars use CSS3 transitions and animations to achieve some of their effects. Internet Explorer 9 and below, as well as older versions of Firefox, do not support these properties. Also keep in mind that Opera 12 does not support animations. The progress bars will still display in these browsers—they just won't be as pretty.

Creating a Progress Bar

To build a progress bar, you create a container `<div>` with the class `.progress`. This creates the full, but empty, bar. Inside that put another `<div>` with the class `.progress-bar`. To be accessible, it also should use the `role="progressbar"`, aria-valuenow, aria-valuemin, and aria-valuemax attributes. Set the aria-valuenow to the current value, aria-valuemin to the minimum possible value, and aria-valuemax to the maximum possible value. Finally, define the size of the progress bar with a style `width` property set to the percent the bar value should be. Listing 18.9 shows the HTML for Figure 18.5—a basic progress bar.

LISTING 18.9 A Basic Progress Bar

```
<div class="progress">
  <div class="progress-bar" role="progressbar" aria-valuenow="20"
      aria-valuemin="0" aria-valuemax="100" style="width: 20%;">
    20% Read
  </div>
</div>
```

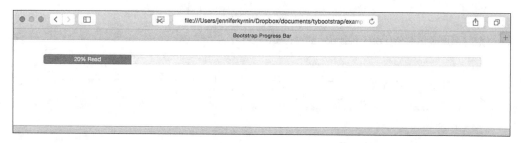

FIGURE 18.5
A basic progress bar.

CAUTION

Leave Space for the Label

If your progress bar is showing a very low progress level, it might not display enough to show your label. To solve this, you should set a `min-width` on the `.progress-bar` element—for example:

```
style="width: 1%; min-width: 2em;"
```

If you don't want a label on your bar, you can surround the text with an `.sr-only` `<span>` tag. This will ensure that screen readers can still evaluate the progress, such as in Listing 18.10.

LISTING 18.10 Progress Bar with No Visible Label

```
<div class="progress">
  <div class="progress-bar" role="progressbar" aria-valuenow="20"
      aria-valuemin="0" aria-valuemax="100" style="width: 20%;">
    <span class="sr-only">20% Read</span>
  </div>
</div>
```

Styling a Progress Bar

The default progress bar is blue, but you can add other colors with contextual classes. As with all contextual classes, if the colors have a meaning, you should indicate that meaning via text somewhere on or near the progress bar. If you don't want that text to display in web browsers, surround it with the .sr-only class. The contextual classes are

▶ .progress-bar-success

▶ .progress-bar-info

▶ .progress-bar-warning

▶ .progress-bar-danger

Figure 18.6 shows the five progress bar styles you can use.

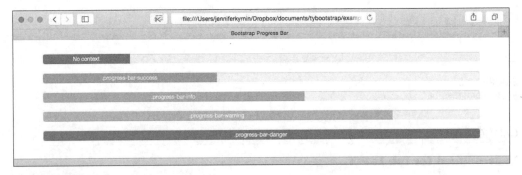

FIGURE 18.6
Five different progress bars.

You also can make your progress bars look even fancier with the .progress-bar-striped class. Plus, if you add the .active class to a striped progress bar, it animates the stripes so they move from right to left. These, however, will not be animated in Internet Explorer 9. If you want all the bars inside a .progress container to be striped, you can use the .progress-striped class on the container element.

If you include several progress bars inside your .progress container, you can stack them to provide additional information. Just make sure the total width of all the bars is not more than 100%.

Create a Bar Graph

You commonly see bar graphs on computers detailing information like the amount of space used by various file types. With the Bootstrap progress bar, you can create a bar graph for your website:

1. Open your Bootstrap web page in a web editor.

2. Add a `.progress` container `<div>`:

   ```
   <div class="progress">
   </div>
   ```

3. Place your first progress bar element inside the `.progress` container:

   ```
   <div class="progress-bar">
     54% love dandelions
   </div>
   ```

4. Add the appropriate contextual class to get the color you want, such as `.progress-bar-info`.

5. Add `.progress-bar-striped` and `.active` to create an animated striped bar.

6. Add the width of the bar in a CSS style. The final HTML will look like this:

   ```
   <div
   class="progress-bar progress-bar-info progress-bar-striped active"
   style="width: 54%">
     54% love dandelions
   </div>
   ```

7. Repeat steps 3–6 for the other bars in your graph.

Figure 18.7 shows what this might look like on a page, and Listing 18.11 shows the HTML for the progress bar.

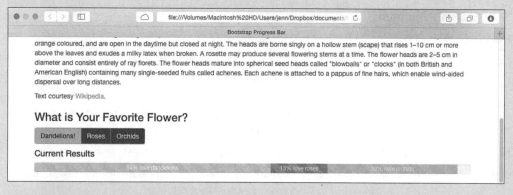

FIGURE 18.7
A stacked progress bar.

LISTING 18.11 A Stacked Progress Bar

```
<div class="progress">
  <div class="progress-bar progress-bar-warning
  progress-bar-striped active" style="width: 54%">
    54% love dandelions
  </div>
  <div class="progress-bar progress-bar-danger progress-bar-striped
  active" style="width: 13%">
    13% love roses
  </div>
  <div class="progress-bar progress-bar-info progress-bar-striped
  active" style="width: 30%">
    30% love orchids
  </div>
</div>
```

Summary

In this hour, you learned how Bootstrap works with CSS3 transitions to create a more robust animation engine for your websites. You also learned how to modify buttons to change their states and toggle on and off and how to turn checkbox and radio input controls into buttons. This hour covered how to make alerts and how to make them dismissible with a click, and you learned how to show the progress of a script or long document with a progress bar as well as how to create a bar graph.

A number of methods and events were covered in this hour. Table 18.1 lists the methods covered, and Table 18.2 lists the events. Table 18.3 lists the CSS classes you learned this hour.

TABLE 18.1 Methods

Plugin	Method	Description
Alert	`$().alert()`	Initializes the alert and sets it to listen for click events on descendant elements to give the element close functionality.
Alert	`$().alert('close')`	Closes the alert by removing it from the DOM. If CSS animations are set, it fades out the alert.
Button	`$().button('reset')`	Swaps the text to the original text, resetting the button.
Button	`$().button(string)`	Swaps the button text to any data-defined text, using a `data-string-text` attribute.
Button	`$().button('toggle')`	Switches the button's state from on to off and back.

TABLE 18.2 Events

Plugin	Event	Description
Alert	close.bs.alert	This fires immediately when the close method is called.
Alert	closed.bs.alert	This fires after the alert is completely closed, including any CSS animations.

TABLE 18.3 CSS Classes

Class	Description
.alert	Defines the container as an alert.
.alert-danger	Indicates that the alert defines something dangerous. Makes the background color red.
.alert-dismissable	Alternative spelling. Indicates that the alert is dismissible and will have a close button inside.
.alert-dismissible	Indicates that the alert is dismissible and will have a close button inside.
.alert-info	Indicates that the alert defines information. Makes the background color light blue.
.alert-link	Matches link color inside an alert to the contextual class of that alert.
.alert-success	Indicates that the alert defines a successful state. Makes the background color green.
.alert-warning	Indicates that the alert defines a warning. Makes the background color yellow.
.progress	Defines the container as a progress bar element.
.progress-bar	Defines the element as a bar inside a progress bar. Turns the bar dark blue.
.progress-bar-danger	Indicates that the bar defines something dangerous. Turns the bar red.
.progress-bar-info	Indicates that the bar defines information. Turns the bar light blue.
.progress-bar-striped	Gives the bar a striped pattern.
.progress-bar-success	Indicates the bar defines a successful state. Turns the bar green.
.progress-bar-warning	Indicates the bar defines a warning state. Turns the bar yellow.
.progress-striped	Gives all bars in this progress bar a striped pattern.

Workshop

The workshop contains quiz questions to help you process what you've learned in this hour. Try to answer all the questions before you read the answers.

Q&A

Q. In your example buttons image, you created a checkbox image that had the same colors as your site design. How did you do that?

A. I added a style sheet that included styles for the `.btn-info` class. Because my styles load after the Bootstrap styles, they override any default styles. Listing 18.12 shows the CSS I used. The second property is more complex because I wanted to catch every possible instance.

LISTING 18.12   CSS to Modify Buttons

```
.btn-info {
  background-color: #025301;
  border-color: #999;
}
.btn-info:hover,  .btn-info:focus,  .btn-info.focus,
.btn-info:active,  .btn-info.active,
.open>.dropdown-toggle.btn-info {
  background-color: #80D464;
  border-color: #999;
  color: #000;
}
```

Q. When should I use alerts on my pages?

A. You can use alerts whenever you want to convey additional information to your customers. But they are best used when something has happened or changed that you need to report to your customers. Some common uses for alerts include

▶ Error messages, such as wrong username or password on a login form

▶ Information about a form field, such as valid characters or length

▶ Welcome messages, such as when a visitor comes to your site from a search engine

▶ Success messages, such as when a file download is complete

Q. I want my progress bar to update automatically based on a timer or some other function. How do I do that?

A. There are many ways to do that. One way is to use jQuery to change the width of the bar periodically. To make it easy, you should add an `id` attribute to the bar you'll be updating. Then you can select that element with jQuery and change the width. For example, the script in Listing 18.13 changes the width of a bar named `#myBar` to 25% and changes the text label as well when the bar is clicked on.

LISTING 18.13 Change the Width of a Progress Bar

```
<script>
$(document).ready(function() {
  $("#myBar").click(function() {
    $(this).css('width','25%');
    $(this).text('25%');
  });
});
</script>
```

Quiz

1. What does the `transition.js` plugin do in the Bootstrap JavaScript?

    a. Gives Bootstrap support to create transitions on links

    b. Helps emulate CSS transitions in browsers that don't support them

    c. Replaces CSS transitions with JavaScript

    d. Both A and B

2. What does the `data-loading-text` attribute do?

    a. Changes the button state.

    b. Loads new button on the page.

    c. Loads text in place of a button.

    d. Nothing; this is not a valid attribute.

3. Which attribute is required to create a pre-toggled button?

    a. The `.active` class

    b. `autocomplete="true"`

    c. `data-toggle="button"`

    d. A and B

    e. A and C

4. What, if anything, is wrong with this HTML?

    ```
    <div class="btn-group">
      <label class="btn btn-primary active">
        <input type="radio" name="flowers" id="dandelion"
               autocomplete="off" checked data-toggle="buttons">
        Dandelions!
      </label>
      <label class="btn btn-primary">
    ```

```
    <input type="radio" name="flowers" id="rose"
          autocomplete="off" data-toggle="buttons">Roses
  </label>
  <label class="btn btn-primary">
    <input type="radio" name="flowers" id="orchid"
          autocomplete="off" data-toggle="buttons">Orchids
  </label>
</div>
```

   a. The `autocomplete` attribute should be set to `on`.

   b. The `.active` class should be placed on the `<input>` tag.

   c. The `data-toggle="buttons"` attribute belongs on the `.btn-group` element.

   d. All of the above.

   e. None of the above

5. True or False: Contextual classes are required for Bootstrap alerts.

6. Which class do you use to indicate an alert can be closed?

   a. `.alert-dismissable`

   b. `.dismiss`

   c. `.dismissible`

   d. `.dismissible-alert`

7. Is the `alert.js` plugin required to add an alert to a Bootstrap page?

   a. Yes, you must include it to add an alert.

   b. Yes, but you can include it using the `bootstrap.js` file.

   c. No, but your alerts won't fade in or out.

   d. No, you require the plugin only if the alert is dismissible.

8. Which plugin is required to add progress bars?

   a. `progress.js`

   b. `progress-bar.js`

   c. `bars.js`

   d. None

9. True or False: Contextual classes are required for Bootstrap progress bars.

10. What does the `.progress-bar-striped` class do?

    a. Styles the bar with stripes.

    b. Styles the bar with animated stripes.

    c. Styles all bars in the progress bar with stripes.

    d. Nothing; the correct class is `.progress-bar-stripe`.

Quiz Answers

1. b. Helps emulate CSS transitions in browsers that don't support them.

2. a. Changes the button state.

3. e. Both `data-toggle="button"` and the `.active` class.

4. c. The `data-toggle="buttons"` attribute belongs on the `.btn-group` element.

5. True

6. a. `.alert-dismissable`

7. d. No, you require the plugin only if the alert is dismissible.

8. d. None

9. False

10. a. Styles the bar with stripes.

Exercises

1. Add a checkbox or radio button group to your web page. Test the page in a browser to see how the buttons remain selected in a checkbox group and how the alternate buttons toggle off in a radio group.

2. Place a dismissible alert somewhere on your page. Include the `.fade` and `.in` classes so that it fades out when it is closed. Don't forget to include a close button.

HOUR 19
Collapse and Accordion

What You'll Learn in This Hour:

▶ How to create a collapsible section
▶ Using the Collapse plugin to build an accordion
▶ How to create an accordion menu
▶ How to hack Bootstrap to build a horizontal collapsible section

In this hour you learn how to create collapsible sections on your website using the Collapse plugin. This will enable you to create sections that can be shown or hidden on the click of a link or button.

One common way that the Collapse plugin is used is to create a website accordion. In this hour you learn how to create accordions to manage your website content, or even build the entire page.

You also learn how to create accordion menus and how to create blocks that collapse horizontally rather than vertically.

The Collapse Plugin

Just like every other Bootstrap plugin, you get the collapsing functions automatically if you include the full `bootstrap.js` file in your pages, or you can install just `collapse.js`. As usual, make sure you have jQuery included in the page as well. If you only install `collapse.js`, you need to install the Transitions plugin (refer to Hour 18, "Transitions, Buttons, Alerts, and Progress Bars") as well.

Creating a Collapsible Section

First, add an element with an ID that you want to show and hide. This is usually a `<div>` but can be any block-level element. Give that element the `.collapse` class. After you place that class on the element, it will disappear from the page. To bring it back, you need to add a button with the `data-toggle="collapse"` attribute and the `data-target` attribute pointing to the ID. If your button is built with an `<a>` tag, you can target the collapse with the `href` attribute. Listing 19.1 shows how this would look. Figure 19.1 shows a collapsed element being opened.

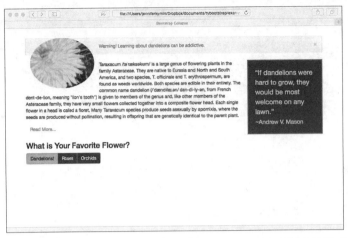

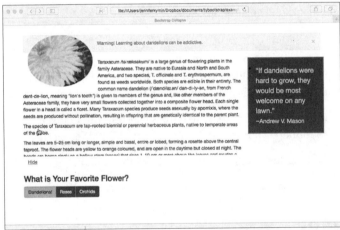

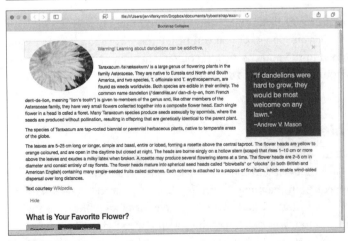

FIGURE 19.1
A collapsed element before (top), during (middle), and after (bottom) clicking on "Read More…".

LISTING 19.1 A Basic Collapsed Element

```
<a class="btn btn-default" data-toggle="collapse"
   href="#myCollapsedSection" aria-expanded="false"
   aria-controls="myCollapsedSection">
  Show the Content
</a>
<article id="myCollapsedSection" class="collapse">
<h2>My Collapsed Section</h2>
<p>Lorem ipsum dolor sit amet, consectetur adipiscing elit. Integer
nec odio. Praesent libero. Sed cursus ante dapibus diam. Sed nisi.
Nulla quis sem at nibh elementum imperdiet. Duis sagittis ipsum.
Praesent mauris. Fusce nec tellus sed augue semper porta. Mauris
massa. Vestibulum lacinia arcu eget nulla. Class aptent taciti
sociosqu ad litora torquent per conubia nostra, per inceptos
himenaeos. Curabitur sodales ligula in libero. </p>
</article>
```

Because the Collapse plugin works on the IDs, you can collapse almost anything as long as it has a unique ID. You also can create multiple buttons that toggle the same element.

CAUTION

You Can't Collapse Paragraphs or Table Cells

In my testing, I found that you cannot collapse paragraphs (<p>) or table cells (<td>, <th>, and so on). To be strictly accurate, I could set the .collapse class on them and they would disappear, but I could not get them to reappear using a data-target="collapse" trigger. If you're planning on collapsing elements other than <div>, <section>, <article>, or other HTML5 sectioning elements, you should test that it works correctly before you launch.

TRY IT YOURSELF ▼

Create a Toggle Link for a Long Blog Post

Many blogs with long articles include a "more" link after a few lines that points to another page with the rest of the article. But with the Collapse plugin and the Button plugin (that you learned about in Hour 18), you can include all the content on the same page and have it toggle on and off. In this Try It Yourself, you will learn how to build this feature:

1. Write the full blog post in your Bootstrap page.

2. Surround the section to be hidden with a <section> tag with a unique ID:

   ```
   <section id="moreDandelions"> ... </section>
   ```

3. Place the .collapse class on the hidden section:

   ```
   <section id="moreDandelions" class="collapse"> ... </section>
   ```

4. Add a read more link, and make sure it has a unique ID and the `href` points to the collapsed section:

```
<a href="#moreDandelions" class="btn btn-link"
    id="moreLink">Read More...</a>
```

5. Add the `data-toggle="collapse"` attribute to the link.

   If you leave it like this, the "Read More..." link will remain on the page as it is. If you click it again, the collapsing section will hide. But that is confusing, so I recommend adding a small jQuery script to toggle the text.

6. Add the following simple script to change the text of the button when it's clicked:

```
<script>
$(document).ready( function() {
  $('a#moreLink').click(function() {
    $(this).text($(this).text() ==
               'Read More...' ? 'Hide' : 'Read More...');
  });
});
</script>
```

With this type of coding on a Bootstrap blog, you don't have to drive readers to another page; they can show and hide the articles they are interested in immediately. Listing 19.2 shows the HTML I used.

LISTING 19.2   HTML to Show and Hide a Blog Post

```
<article>
<p>Taraxacum /təˈræksəkʉm/ is a large genus of flowering plants in
the family Asteraceae. They are native to Eurasia and North and
South America, and two species, T. officinale and T.
erythrospermum, are found as weeds worldwide. Both species are
edible in their entirety. The common name dandelion (/ˈdændɨlaɪ.ən/
dan-di-ly-ən, from French dent-de-lion, meaning "lion's tooth") is
given to members of the genus and, like other members of the
Asteraceae family, they have very small flowers collected together
into a composite flower head. Each single flower in a head is
called a floret. Many Taraxacum species produce seeds asexually by
apomixis, where the seeds are produced without pollination,
resulting in offspring that are genetically identical to the parent
plant.</p>
<section id="moreDandelions" class="collapse">
<p>The species of Taraxacum are tap-rooted biennial or perennial
herbaceous plants, native to temperate areas of the globe.</p>
<p>The leaves are 5-25 cm long or longer, simple and basal, entire
or lobed, forming a rosette above the central taproot. The flower
heads are yellow to orange coloured, and are open in the daytime
```

```
but closed at night. The heads are borne singly on a hollow stem
(scape) that rises 1-10 cm or more above the leaves and exudes a
milky latex when broken. A rosette may produce several flowering
stems at a time. The flower heads are 2-5 cm in diameter and
consist entirely of ray florets. The flower heads mature into
spherical seed heads called "blowballs" or "clocks" (in both
British and American English) containing many single-seeded fruits
called achenes. Each achene is attached to a pappus of fine hairs,
which enable wind-aided dispersal over long distances.</p>
<p>Text courtesy <a href="http://en.wikipedia.org/wiki/Taraxacum">
Wikipedia</a>.</p>
</section>
<p><a href="#moreDandelions" class="btn btn-link" id="moreLink"
    data-toggle="collapse" aria-expanded="false"
    aria-controls="moreDandelions">Read More...</a></p>
</article>

<script src="http://code.jquery.com/jquery-latest.js"></script>
<script src="js/bootstrap.min.js"></script>
<script>
$(document).ready( function() {
  $('a#moreLink').click(function() {
    $(this).text($(this).text() ==
    'Read More...' ? 'Hide' : 'Read More...');
  });
});
</script>
```

When you create a collapsible section, Bootstrap uses three classes to create the change:

- ▶ `.collapse`—This hides the content.

- ▶ `.collapse.in`—These two classes combined show the content. So if you want your collapsible element to start open, use `class="collapse in"`.

- ▶ `.collapsing`—This is applied during the transitions between showing and hiding the content.

If you want to collapse a section of your page after a JavaScript event, you can apply these classes yourself or use the methods provided in the `collapse.js` plugin.

To create the button or link that controls the collapsible object, you use the `data-toggle="collapse"` and `data-target` attributes. The `data-target` should point to a CSS selector of the collapsible object.

Horizontal Collapsing Elements

The default collapse in Bootstrap is vertical. The collapsing element slides in from top to bottom, pushing any content below it lower down on the page. But, with some extra CSS, you can make your Bootstrap elements collapse horizontally or from right to left.

Bootstrap has support for the .width class within the collapse.js plugin. By placing this class on the collapsed element, you can change the behavior from a vertical collapse to a horizontal collapse.

However, Bootstrap does not have the corresponding transitions in the style sheet, so you will need to add them. Listing 19.3 shows the CSS you should add to your style sheet.

LISTING 19.3 Adding Horizontal Transitions to Collapse

```
.collapsing.width {
  -webkit-transition-property: width, visibility;
  transition-property: width, visibility;
  width: 0;
  min-height: 100px;
}
```

So, to collapse a section horizontally, you should add the .width class to the .collapse element. Then, with the CSS in Listing 19.3, your element will collapse horizontally rather than vertically.

If you have collapsed a block of text, you might notice that the bottom is cut off while the text is loading. This is caused by the min-height property. If you would like more text to display, make that value larger. If you're sliding images inside your collapse, you should make the min-height the height of your largest image.

Collapse Options

The two options you can use with the collapse.js plugin are

▶ parent—This is a Boolean (true/false) option that takes the selector of the parent element to create a style of accordion out of the collapsible elements. The default value is true.

▶ toggle—This is a Boolean option that toggles the collapsible element on invocation. The default value is true.

CAUTION

The toggle Option Applies Only on Invocation

Be aware that you cannot apply the toggle option more than once on a collapsible element. If you need the element to be toggled, you must use the appropriate collapse method instead.

Collapse Methods

The Collapse plugin has four methods you can call in your scripts:

- ▶ `$().collapse(options)`—This method activates the content as a collapsible element. The options object is optional.
- ▶ `.collapse('toggle')`—This toggles a collapsible element to be shown or hidden.
- ▶ `.collapse('show')`—This shows a collapsible element.
- ▶ `.collapse('hide')`—This hides a collapsible element.

Collapse Events

The four events you can watch for in the `collapse.js` plugin are

- ▶ `show.bs.collapse`—This fires immediately when the `show()` method is called.
- ▶ `shown.bs.collapse`—This fires when the collapsed element has been made visible to the user, waiting for any CSS transitions to complete.
- ▶ `hide.bs.collapse`—This fires immediately when the `hide()` method is called.
- ▶ `hidden.bs.collapse`—This fires when the collapsed element has been completely hidden from the user, waiting for any CSS transitions to complete.

Accordions

Accordions are multiple collapsible sections that close automatically when other parts of the accordion are opened. The Bootstrap Collapse plugin lets you create them easily with the `data-parent` attribute. Figure 19.2 shows an accordion on a web page.

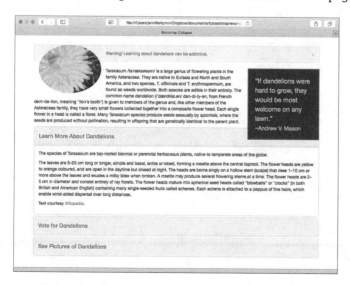

FIGURE 19.2
An accordion on a web page.

Creating Accordions

You can create accordions in many ways, but the easiest is with panel groups and list groups (refer to Hour 12, "Creating Navigation Systems with Bootstrap"). These work best because they provide easily recognizable headings and content areas that can be shown and hidden.

▼ TRY IT YOURSELF

Create an Accordion from a Panel Group

Accordions use a lot of HTML to create the hidden and unhidden content. In this Try It Yourself, you learn the steps to turn any panel group into an accordion using the Collapse plugin and the `.panel-collapse` class:

1. Open your Bootstrap page in an HTML editor, and create a panel with a panel heading. Make sure your `.panel-heading` has a unique ID:

   ```
   <div class="panel panel-default">
     <div class="panel-heading" role="tab" id="moreDandy">
       <h4 class="panel-title">Learn More About Dandelions</h4>
     </div>
     <div class="panel-body">
       <p>The species of Taraxacum are tap-rooted biennial or
       perennial herbaceous plants, native to temperate areas of the
       globe.</p>
       <p>...</p>
       <p>Text courtesy
       <a href="http://en.wikipedia.org/wiki/Taraxacum">Wikipedia</a>.
       </p>
     </div>
   </div>
   ```

2. Surround the `.panel-body` with another `<div>` that has the `.panel-collapse` class on it:

   ```
   <div id="moreDandelions" class="panel-collapse "
       role="tabpanel" aria-labelledby="moreDandy">
     <div class="panel-body">
   ...
   ```

3. Add the class `.collapse` to the `.panel-collapse` element, and if you want that panel to display immediately, add the `.in` class.

4. Link your `.panel-title` to create the trigger:

   ```
   <h4 class="panel-title">
     <a href="#moreDandelions" aria-expanded="true"
       aria-controls="moreDandelions">Learn More About
       Dandelions</a>
   </h4>
   ```

5. Add the `data-toggle="collapse"` attribute to the link to turn it into a trigger.

6. Add a `data-parent="#myAccordion"` attribute to the link. This will point to the entire panel group, which is created in step 7.

7. Surround the entire panel with a `.panel-group` element. This can be any block-level tags, but I prefer `<section>` or `<div>`. Make sure that this has the same ID that you pointed to in step 6:

```
<section class="panel-group" role="tablist"
    aria-multiselectable="true" id="myAccordion">
```

8. Follow steps 1–6 to create additional panels for your accordion. Remember to remove the `.in` class from any you don't want displaying on the initial load.

Accordions can seem complicated, but if you approach them systematically, you will find they are easy to build. Listing 19.4 shows the full HTML for a working three-panel accordion.

LISTING 19.4   A Three-Panel Accordion

```
<section class="panel-group" role="tablist"
        aria-multiselectable="true" id="accordion">
  <div class="panel panel-default">
    <div class="panel-heading" role="tab" id="moreDandy">
      <h4 class="panel-title">
        <a data-toggle="collapse" data-parent="#accordion"
          href="#moreDandelions" aria-expanded="true"
          aria-controls="moreDandelions">Learn More About
          Dandelions</a>
      </h4>
    </div>
    <div id="moreDandelions" class="panel-collapse collapse in"
        role="tabpanel" aria-labelledby="moreDandy">
      <div class="panel-body">
        <p>The species of Taraxacum are tap-rooted biennial or
        perennial herbaceous plants, native to temperate areas of the
        globe.</p>
        <p>...</p>
        <p>Text courtesy
        <a href="http://en.wikipedia.org/wiki/Taraxacum">Wikipedia</a>.
        </p>
      </div>
    </div>
  </div>
  <div class="panel panel-default">
    <div class="panel-heading" role="tab" id="voteDandy">
      <h4>
```

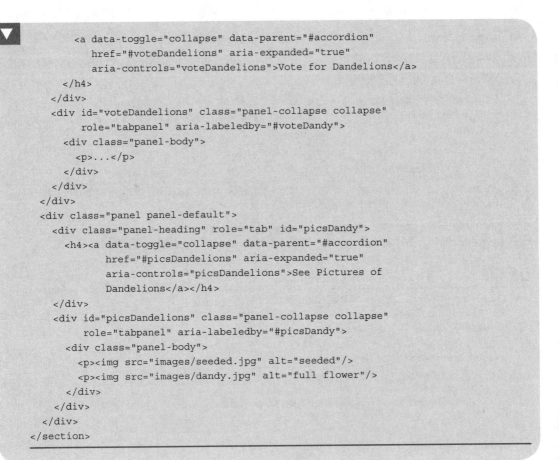

```
               <a data-toggle="collapse" data-parent="#accordion"
                  href="#voteDandelions" aria-expanded="true"
                  aria-controls="voteDandelions">Vote for Dandelions</a>
          </h4>
        </div>
        <div id="voteDandelions" class="panel-collapse collapse"
             role="tabpanel" aria-labeledby="#voteDandy">
          <div class="panel-body">
            <p>...</p>
          </div>
        </div>
      </div>
      <div class="panel panel-default">
        <div class="panel-heading" role="tab" id="picsDandy">
          <h4><a data-toggle="collapse" data-parent="#accordion"
                 href="#picsDandelions" aria-expanded="true"
                 aria-controls="picsDandelions">See Pictures of
                 Dandelions</a></h4>
        </div>
        <div id="picsDandelions" class="panel-collapse collapse"
             role="tabpanel" aria-labeledby="#picsDandy">
          <div class="panel-body">
            <p><img src="images/seeded.jpg" alt="seeded"/>
            <p><img src="images/dandy.jpg" alt="full flower"/>
          </div>
        </div>
      </div>
    </div>
</section>
```

Using Accordions as Navigation

Accordions are often used to create collapsible navigation on web pages. You can replace the
.panel-body elements in Listing 19.4 with a list group to create a simple menu. Listing 19.5
shows the altered HTML.

LISTING 19.5   List Groups in Accordions

```
<div class="panel-group" id="accordionMenu">
  <div class="panel panel-info">
    <div class="panel-heading">
      <h4 class="panel-title">
        <a data-toggle="collapse" data-parent="#accordionMenu"
           href="#menu1" aria-expanded="true"
           aria-controls="menu1">Menu 1</a>
```

```
      </h4>
    </div>
    <div class="panel-collapse collapse" id="menu1">
      <ul class="list-group">
        <li class="list-group-item"><a href="#">Item 1</a></li>
        <li class="list-group-item"><a href="#">Item 2</a></li>
        <li class="list-group-item"><a href="#">Item 3</a></li>
      </ul>
    </div>
  </div>
  <div class="panel panel-default">
    <div class="panel-heading">
      <h4 class="panel-title">
        <a data-toggle="collapse" data-parent="#accordionMenu"
           href="#menu2" aria-expanded="true"
           aria-controls="menu2">Menu 2</a>
      </h4>
    </div>
    <div class="panel-collapse collapse" id="menu2">
      <ul class="list-group">
        <li class="list-group-item"><a href="#">Item 1</a></li>
        <li class="list-group-item"><a href="#">Item 2</a></li>
        <li class="list-group-item"><a href="#">Item 3</a></li>
      </ul>
    </div>
  </div>
  <div class="panel panel-default">
    <div class="panel-heading">
      <h4 class="panel-title">
        <a data-toggle="collapse" data-parent="#accordionMenu"
           href="#menu3" aria-expanded="true"
           aria-controls="menu3">Menu 3</a>
      </h4>
    </div>
    <div class="panel-collapse collapse" id="menu3">
      <ul class="list-group">
        <li class="list-group-item"><a href="#">Item 1</a></li>
        <li class="list-group-item"><a href="#">Item 2</a></li>
        <li class="list-group-item"><a href="#">Item 3</a></li>
      </ul>
    </div>
  </div>
</div>
```

This creates a panel group with list groups inside it instead of `.panel-body` elements.
But if you want to have a more compact menu, you can take a few shortcuts. Create your
menu as one large list group using <a> tags. Surround the submenus with separate <div

class="collapse"> tags. Give each submenu a unique ID, and then create a .list-group-item above them as the trigger to open them. In the <div class="list-group"> tag, add the .panel class. Then surround the entire list group with a container that will act as your data-parent and assign that to all your triggers. Listing 19.6 shows how this would look, and Figure19.3 shows the menu in a web page.

LISTING 19.6   A Modified List Group Accordion

```
<div id="menu">
  <div class="panel list-group" role="tablist"
      aria-multiselectable="true">
    <a href="#" class="list-group-item" data-toggle="collapse"
       data-target="#dandyDeets" data-parent="#menu"
       aria-expanded="false"
       aria-controls="#dandyDeets">DETAILS</a>
    <div id="dandyDeets" class="collapse submenu">
      <a class="list-group-item small">taxonomy</a>
      <a class="list-group-item small">colors</a>
      <a class="list-group-item small">sizes</a>
    </div>
    <a href="#" class="list-group-item" data-toggle="collapse"
       data-target="#dandyArts" data-parent="#menu"
       aria-expanded="false"
       aria-controls="#dandyArts">ARTICLES</a>
    <div id="dandyArts" class="collapse submenu">
      <a class="list-group-item small">Dandelions and You</a>
      <a class="list-group-item small">Dandelion's Best Friend</a>
      <a class="list-group-item small">Read More</a>
    </div>
    <a href="#" class="list-group-item" data-toggle="collapse"
       data-target="#dandyRecipes" data-parent="#menu"
       aria-expanded="false"
       aria-controls="#dandyRecipes">RECIPES</a>
    <div id="dandyRecipes" class="collapse submenu">
      <a class="list-group-item small">Dandelion Soup</a>
      <a class="list-group-item small">Dandelion Wine</a>
      <a class="list-group-item small">Search Recipes</a>
    </div>
  </div>
</div>
```

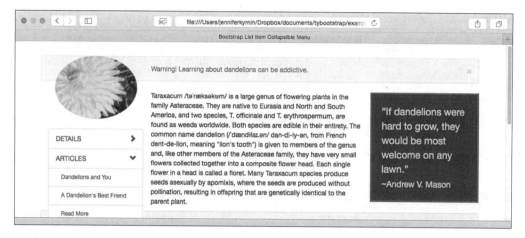

FIGURE 19.3
A simple collapsible navigation menu.

I added a Glyphicon `<span class="pull-right glyphicon glyphicon-chevron-right"></span>` to the menu in Figure 19.3 to make them show up more clearly.

Summary

In this hour you learned how to use the Bootstrap `collapse.js` plugin to create collapsible sections and accordions on your web pages. You learned the options (shown in Table 19.1), methods (shown in Table 19.2), and events (shown in Table 19.3) you can use to add collapsing sections to your scripts. You also learned the CSS classes Bootstrap uses (shown in Table 19.4) to create the collapsible sections.

TABLE 19.1   Collapse Options

Option	Description
parent	All collapsible elements under the same parent will be closed when this element is opened. It requires the `.panel` class and creates an accordion.
toggle	Toggles the collapsible element when this option is first invoked.

TABLE 19.2   Collapse Methods

Method	Description
`$().collapse(options)`	Activates the element as collapsible content. It will accept optional options as an object.
`.collapse('hide')`	Hides a collapsible element.
`.collapse('show')`	Shows a collapsible element.
`.collapse('toggle')`	Toggles a collapsible element to shown or hidden.

TABLE 19.3   Collapse Events

Event	Description
`hide.bs.collapse`	Fires immediately as the element is hidden.
`hidden.bs.collapse`	Fires after the element is completely hidden, including any CSS transitions.
`show.bs.collapse`	Fires immediately as the element is shown.
`shown.bs.collapse`	Fires after the element is completely visible, including after any CSS transitions.

TABLE 19.4   CSS Classes

Class	Description
`.collapse`	Hides the element and indicates that it is collapsible.
`.collapsing`	Placed on the element as it is collapsing (both during hide and show) to add transitions.
`.panel-collapse`	Indicates a panel element that is collapsible. Used in accordions.
`.panel-group`	Indicates a group of panel elements used to create an accordion.
`.width`	A hidden class used to create horizontal collapsing elements.

Workshop

The workshop contains quiz questions to help you process what you've learned in this hour. Try to answer all the questions before you read the answers.

Q&A

Q. Why didn't you use the Bootstrap `button.js` plugin to change the state of the button in the blog post Try It Yourself?

A. One of the drawbacks to using Bootstrap plugins is that you can only use one plugin on an element at a time. In situations where you want to, you need to add a container element around the first and apply the second plugin to that element. Or you can do like I did and write your own JavaScript to take care of the problem.

Q. I'm updating an older site that uses Bootstrap, and the accordions on it don't work in Internet Explorer 9. What can I do?

A. Chances are good that this site is using an older version of Bootstrap. Bootstrap 3 accordions will work in Internet Explorer 9. If you don't remember how to get Bootstrap 3, that was covered in Hour 2, "Downloading and Installing Bootstrap." Hour 3, "Build Your First Bootstrap Website with the Basic Template," will help you verify what version your site is using.

Q. I wanted to add an icon to my headings to indicate whether the element is open or closed. But when I add it, it never changes. How do I fix that?

A. You have to add your own custom script. You can modify Listing 19.6 to add `<span class="pull-right glyphicon glyphicon-chevron-right"></span>` to each of the `data-toggle` triggers. Then you can add the JavaScript from Listing 19.7 to switch the right chevron to a down chevron.

LISTING 19.7   JavaScript to Toggle Menu Icons

```
$(document).ready(function(){
  $('#menu .collapse').on('show.bs.collapse', function() {
    var $head = $(this).prev();
    $head.find('span').removeClass('glyphicon-chevron-right')
                      .addClass('glyphicon-chevron-down');
  });
  $('#menu .collapse').on('hide.bs.collapse', function() {
    var $head = $(this).prev();
    $head.find('span').removeClass('glyphicon-chevron-down')
                      .addClass('glyphicon-chevron-right');
  });
});
```

Quiz

1. True or False: To create a collapsible section, all you need is the `collapse.js` plugin.

2. How can you make a collapsed section visible on the first load?

    a. You don't need to do anything because they are visible on first load.

    b. Add the class `.visible` to the element.

    c. Add the `data-visibility` attribute to the trigger.

    d. Add the class `.in` to the element.

3. What is wrong with this trigger element HTML code?

    ```
    <button class="btn btn-default" data-toggle="collapse"
            data-target="#myCollapsedSection">
      Show Me
    </button>
    ```

    a. It needs an `href` attribute to point to the collapsible element.

    b. It requires the `aria-controls` and `aria-expanded` attributes.

    c. It must be an `<a>` tag.

    d. Nothing; this should work fine.

4. How can you change the button text after it has triggered a collapse?

    a. Add the state change in the `button.js` plugin to the button.

    b. Use the attribute `data-collapsed` to set the text.

    c. You must write custom JavaScript.

    d. You can't.

5. Which of the following is not a class added by the Collapse plugin?

    a. `.active`

    b. `.collapse`

    c. `.collapsing`

    d. `.collapse.in`

6. Does Bootstrap allow for elements to collapse horizontally?

    a. Yes, it's built in using the `.width` class.

    b. Yes, it's built in using the `.width` class, but you have to write your own transitions.

    c. No, you have to add support for the `.width` class to the `bootstrap.js` file.

    d. No, because it's impossible to collapse elements horizontally.

7. What does the `parent` option do?

    a. It defines the trigger element of a collapsible section.

    b. It defines the collapsible element of a collapsible section.

    c. It defines the container element of an accordion.

    d. It defines the collapsible element of an accordion.

8. True or False: An accordion will close an open section when opening a new section.

9. True or False: An accordion must have the `#accordion` ID to work.

10. Which of the following can be used for accordion menus?

    a. Panel groups

    b. List groups

    c. Tables

    d. A and B

    e. All of the above

Quiz Answers

1. False. You also need the transitions plugin and jQuery.

2. d. Add the class `.in` to the element.

3. d. Nothing; this should work fine.

4. c. You must write custom JavaScript.

5. a. `.active`

6. b. Yes, it's built in using the `.width` class, but you have to write your own transitions.

7. c. It defines the container element of an accordion.

8. True. That is what defines an accordion.

9. False. I use that ID for convenience, not because it's required in the code.

10. b. List groups.

Exercises

1. Add a collapsible section to your web page.

2. Add an accordion to your web page. Try to add some icons to the trigger elements, and see whether you can toggle them up and down (refer to the Q&A section for more help).

Carousels

What You'll Learn in This Hour:

▶ How to create an image slideshow on a Bootstrap site

▶ How to use the `carousel.js` plugin

▶ Best practices for using carousels on websites

▶ Common problems with carousels and how to deal with them

Web carousels are a popular way to create image slideshows and display multiple marketing messages on a web page without taking up as much vertical space. This allows the images or marketing to stay "above the fold" and engage more customers.

But carousels can be notoriously difficult to build and maintain. You have to know a lot of JavaScript, and often the scripts you can use require a lot of editing just to add another slide. Bootstrap makes this easy with a built-in plugin—`carousel.js`—that you can use to create creative and interesting slideshows on your web pages.

Creating Carousels

Bootstrap makes carousels easy by including the `carousel.js` file in your HTML along with jQuery. If you include the `bootstrap.js` file, you'll have carousel support built in.

Then, to build the carousel, the three sections in the HTML are

▶ **Carousel indicators**—These are a list of the slides in the carousel with pointers to the data. They display the small linked dots at the bottom of the carousel that indicate how many slides there are.

▶ **The slides**—These are the slides. You should have one slide for each indicator.

▶ **The controls**—These are the next and previous links that let readers move from one slide to the next.

To build a carousel, you need to enclose the entire thing in a container element, such as a `<div>`, with the class `.carousel` and a unique ID and the attribute `data-ride="carousel"`:

```
<div id="myCarousel" class="carousel" data-ride="carousel"></div>
```

Inside this `<div>`, create your indicators. The indicators are an ordered list with the `.carousel-indicators` class containing empty list tags that define the slides. Each list item has a `data-target` pointing to the ID of your carousel and a `data-slide-to` attribute with the number of the slide, starting at zero (0). Listing 20.1 shows indicators for a three-slide carousel.

LISTING 20.1   Indicators for a Three-Slide Carousel

```
<!-- Indicators -->
<ol class="carousel-indicators">
  <li data-target="#myCarousel" data-slide-to="0"
      class="active"></li>
  <li data-target="#myCarousel" data-slide-to="1"></li>
  <li data-target="#myCarousel" data-slide-to="2"></li>
</ol>
```

The default color of the indicators is white, so if you are going to slides with a light or white background, especially near the bottom, you should use CSS to color them. Change the background color of the `.carousel-indicators li` and `.carousel-indicators .active` properties.

NOTE

Indicators Are Not Required

If you don't want to have indicators on your carousel, you can leave them out—the carousel will still work. You also can have more indicators than you have slides or more slides than you have indicators. The indicators are just a way to help your customers know that there is more content available than what is currently visible.

After you have your slide indicators, you can add your slides. These are a `<div>` tag with the `.carousel-inner` class containing multiple `.item` containers with the slide images or information. Listing 20.2 shows HTML for three slides.

LISTING 20.2   Three Simple Carousel Slides

```
<!-- Wrapper for slides -->
<div class="carousel-inner" role="listbox">
  <div class="item active">
    <h1>Slide 1</h1>
  </div>
  <div class="item">
    <h1>Slide 2</h1>
  </div>
  <div class="item">
```

```
    <h1>Slide 3</h1>
  </div>
</div>
```

Create a slide for each indicator, and add the `.active` class to the same slide you made active in the indicators list. Make sure that at least one of the items has the `.active` class. Otherwise, your carousel won't display anything.

The last things you need to add are the controls. These are two anchor tags (`<a>`) that use the `.left` and `.right` classes to position them on the left and right of the carousel. They are given the `.carousel-control` class, designating them a carousel control element. The `data-slide` attribute indicates whether the controls should slide the carousel left or right, and the `href` attribute points to the carousel ID. You can put anything inside the anchor tags, but most designers use the Glyphicon chevron icons. Listing 20.3 shows the HTML for a standard set of carousel controls.

LISTING 20.3   Standard Carousel Controls

```
<!-- Controls -->
<a class="left carousel-control" href="#myCarousel" role="button"
   data-slide="prev">
  <span class="glyphicon glyphicon-chevron-left"
        aria-hidden="true"></span>
  <span class="sr-only">Previous</span>
</a>
<a class="right carousel-control" href="#myCarousel" role="button"
   data-slide="next">
  <span class="glyphicon glyphicon-chevron-right"
        aria-hidden="true"></span>
  <span class="sr-only">Next</span>
</a>
```

Like the indicators, the controls are not required, but it is a good idea to use them because they help your customers understand that more content is available. If your carousel doesn't automatically switch between slides, there would be no way to go to another slide without controls or indicators. Figure 20.1 shows a basic carousel using the HTML from Listings 20.1, 20.2, and 20.3.

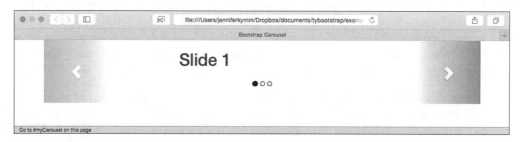

FIGURE 20.1
A basic Bootstrap carousel.

Basic Carousels

Image carousels are arguably the reason carousels were created in the first place. People wanted to be able to create a slideshow on the Web to display many images in a small space. Figure 20.2 shows a typical image carousel. It has some slight modifications to the CSS to center the images (using the `.center-block` class) and change the controls' color (by styling the `.carousel-control.right` and `.carousel-control.left` classes).

FIGURE 20.2
A standard image carousel.

Images are easy to use in carousels. At the most basic level, all you do is replace the `<h1>` tags in Listing 20.2 with `<img>` tags. Then if you are using indicators, make sure that you have the same number of indicators as you have slides and that they point to the correct ones.

However, to create really nice-looking slideshows, you should be aware of these best practices:

▶ The default slide size is 900 × 500 pixels. Image galleries look best when the images are that same ratio, if not the same size.

▶ It's best if all your images are the same size and orientation. The gallery will still work with differently sized images—they just might look a little strange.

▶ The controls default to white icons on a gray gradient. This gradient will overlay images that are 900 × 500, making the edges darker. Be aware of that if you use CSS to change the colors.

▶ The indicators also default to white color. If your images are brightly colored, the indicators might not show up.

One feature of carousels that is really useful on photo galleries is the optional caption. You place another element with the `.carousel-caption` class on it. Then you can place any HTML you want inside the caption, and the text will be written in white just above the indicators. Listing 20.4 shows the `.carousel-inner` element using the HTML5 `<figure>` and `<figcaption>` elements.

LISTING 20.4   Carousel with Captions

```
<div class="carousel-inner" role="listbox">
  <figure class="item active">
    <img src="images/IMG_1958.jpg" alt="dandelion 1"
        class="center-block img-responsive">
    <figcaption class="carousel-caption">
      A Dandy Closeup
    </figcaption>
  </figure>
  <figure class="item">
    <img src="images/IMG_1960.jpg" alt="dandelion 2"
        class="center-block img-responsive">
    <figcaption class="carousel-caption">
      Taking Over the Yard
    </figcaption>
  </figure>
  <figure class="item">
    <img src="images/IMGP1382.jpg" alt="dandelion 3"
        class="center-block img-responsive">
    <figcaption class="carousel-caption">
      One
    </figcaption>
  </figure>
</div>
```

Fancier Carousels

One of the advantages to using Bootstrap carousels is that you're not limited to just images. You can create entire HTML blocks inside the carousel that can then rotate to show off various things.

For example, you might have a series of testimonials you'd like to showcase on your website. These usually consist of a quote, a photo, a link, and a by-line. In other carousel systems, you would have to create these as an image and then link the entire image. But with Bootstrap, you can create a great-looking HTML feature, like in Figure 20.3, and rotate between several. Listing 20.5 shows how you might build this carousel.

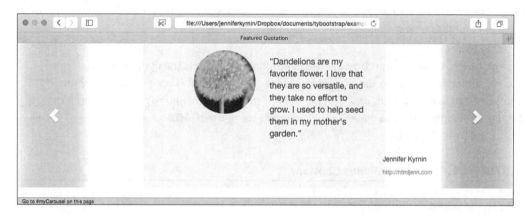

FIGURE 20.3
A carousel of quotations.

LISTING 20.5   A Carousel of Quotations

```
<div id="myCarousel" class="carousel slide" data-ride="carousel">
  <div class="carousel-inner" role="listbox">
    <div class="item active">
      <div class="container-fluid">
        <div class="feature col-md-6 col-md-offset-3">
          <div class="container">
            <div class="row">
              <img src="images/quote1.jpg"
                   class="img-circle col-md-2 col-md-offset-1"
                   alt=""/>
              <blockquote class="col-md-3">
                &#8220;Dandelions are my favorite flower. I love
                that they are so versatile, and they take no effort
                to grow. I used to help seed them in my mother's
                garden.&#8221;
              </blockquote>
            </div>
            <div class="row">
              <p class="col-md-offset-6 col-md-2">Jennifer
              Kyrnin</p>
            </div>
            <div class="row">
              <p class="small col-md-offset-6 col-md-2">
                <a href="http://htmljenn.com/">
```

```
                    http://htmljenn.com</a></p>
            </div>
          </div>
        </div>
      </div>
    </div>

    <div class="item">
      <div class="container-fluid">
        <div class="feature col-md-6 col-md-offset-3">
          <div class="container">
            <div class="row">
              <img src="images/quote2.jpg"
                   class="img-circle col-md-2 col-md-offset-1"
                   alt=""/>
              <blockquote class="col-md-3">
                &#8220;Blah blaherbe blah&#8221;
              </blockquote>
            </div>
            <div class="row">
              <p class="col-md-offset-6 col-md-2">Joseph Blue</p>
            </div>
            <div class="row">
              <p class="small col-md-offset-6 col-md-2">
                <a href="">http://dandylions.weed</a></p>
            </div>
          </div>
        </div>
      </div>
    </div>

    <div class="item">
      <div class="container-fluid">
        <div class="feature col-md-6 col-md-offset-3">
          <div class="container">
            <div class="row">
              <img src="images/quote3.jpg"
                   class="img-circle col-md-2 col-md-offset-1"
                   alt=""/>
              <blockquote class="col-md-3">
                &#8220;Blahbity blah blah&#8221;
              </blockquote>
            </div>
            <div class="row">
              <p class="col-md-offset-6 col-md-2">Josephine
              Blobe</p>
            </div>
            <div class="row">
```

```
            <p class="small col-md-offset-6 col-md-2">
                <a href="">http://dandelyon.plant</a></p>
          </div>
        </div>
      </div>
    </div>
  </div>
</div>
<!-- Controls -->
<a class="left carousel-control" href="#myCarousel" role="button"
   data-slide="prev">
  <span class="glyphicon glyphicon-chevron-left"
        aria-hidden="true"></span>
  <span class="sr-only">Previous</span>
</a>
<a class="right carousel-control" href="#myCarousel"
   role="button" data-slide="next">
  <span class="glyphicon glyphicon-chevron-right"
        aria-hidden="true"></span>
  <span class="sr-only">Next</span>
</a>
</div>
```

Another simple trick to create fancier carousels is to add the `.slide` class to the `.carousel` element. This tells the browser to animate the transition of the slides:

```
<div id="myCarousel" class="carousel slide" data-ride="carousel">
```

Using the Carousel Plugin

The carousel plugin works like every other Bootstrap plugin. You include the `carousel.js` file at the bottom of your HTML and include jQuery. You need the `transition.js` file as well. But be aware that the transitions won't work in Internet Explorer 8 and 9. If you need the slide transitions, you have to create fallback options yourself. If you include the full `bootstrap.js` file, that will include carousel support as well.

If you want your carousel to start animating immediately when the page loads, you must include the `data-ride="carousel"` attribute on the `.carousel` element. But do not use this attribute if you are going to initialize that carousel with JavaScript.

The `data-slide` attribute accepts the keywords `prev` and `next`. This alters the slide position relative to its current position. If you use the `data-slide-to` attribute, you define a specific slide to open based on its index. The index is the number of the slide starting at zero (0).

You can initialize a carousel with JavaScript by writing the following:

```
$('.carousel').carousel()
```

Adding Multiple Carousels

You can have multiple carousels on one page, but each one must have a unique `id` attribute on the `.carousel` element. Be sure that the controls for each carousel point to the correct one. Listing 20.6 shows the HTML for two carousels on the same page.

LISTING 20.6 Two Carousels on One Page

```
<div id="myCarousel" class="carousel" data-ride="carousel">
  <!-- Indicators -->
  <ol class="carousel-indicators">
    <li data-target="#myCarousel" data-slide-to="0"
        class="active"></li>
    <li data-target="#myCarousel" data-slide-to="1"></li>
    <li data-target="#myCarousel" data-slide-to="2"></li>
  </ol>

  <!-- Wrapper for slides -->
  <div class="carousel-inner" role="listbox">
    <div class="item active">
      <h1>Carousel 1 Slide 1</h1>
    </div>
    <div class="item">
      <h1>Carousel 1 Slide 2</h1>
    </div>
    <div class="item">
      <h1>Carousel 1 Slide 3</h1>
    </div>
  </div>

  <!-- Controls -->
  <a class="left carousel-control" href="#myCarousel"
     role="button" data-slide="prev">
    <span class="glyphicon glyphicon-chevron-left"
          aria-hidden="true"></span>
    <span class="sr-only">Previous</span>
  </a>
  <a class="right carousel-control" href="#myCarousel"
     role="button" data-slide="next">
    <span class="glyphicon glyphicon-chevron-right"
          aria-hidden="true"></span>
    <span class="sr-only">Next</span>
  </a>
</div>
<p>and here is some text</p>

<div id="myCarousel2" class="carousel" data-ride="carousel">
  <!-- Indicators -->
```

```
<ol class="carousel-indicators">
  <li data-target="#myCarousel2" data-slide-to="0"
      class="active"></li>
  <li data-target="#myCarousel2" data-slide-to="1"></li>
  <li data-target="#myCarousel2" data-slide-to="2"></li>
</ol>

<!-- Wrapper for slides -->
<div class="carousel-inner" role="listbox">
  <div class="item active">
    <h1>Carousel 2 Slide 1</h1>
  </div>
  <div class="item">
    <h1>Carousel 2 Slide 2</h1>
  </div>
  <div class="item">
    <h1>Carousel 2 Slide 3</h1>
  </div>
</div>

<!-- Controls -->
<a class="left carousel-control" href="#myCarousel2"
   role="button" data-slide="prev">
  <span class="glyphicon glyphicon-chevron-left"
        aria-hidden="true"></span>
  <span class="sr-only">Previous</span>
</a>
<a class="right carousel-control" href="#myCarousel2"
   role="button" data-slide="next">
  <span class="glyphicon glyphicon-chevron-right"
        aria-hidden="true"></span>
  <span class="sr-only">Next</span>
</a>
</div>
```

Carousel Options

The four options you can adjust for your carousels are

▶ interval—This is the amount of time, in milliseconds, that the carousel should pause before cycling to the next slide. The default is 5000; if you use the keyword false, the carousel will not automatically cycle to the next slide.

▶ pause—This turns the pause feature on or off. The default is hover, and that pauses the cycling when the mouse is hovering over the element. Other values will disallow pausing.

- ► `wrap`—This defines whether the cycle should start over at the beginning when the last slide is reached. The default value is `true`.

- ► `keyboard`—This defines whether the carousel should respond to keyboard events. The default value is `true`, and this is the most accessible choice.

Carousel Methods

The six methods you can use to control your carousels are as follows:

- ► `$().carousel(options)`—This method initializes the carousel and starts cycling through the items.

- ► `.carousel('cycle')`—This cycles through the items from left to right.

- ► `.carousel('pause')`—This stops the carousel from cycling through items.

- ► `.carousel(number)`—This cycles the carousel to a specific item, by number, with zero (0) as the first item.

- ► `.carousel('prev')`—This cycles to the previous item.

- ► `.carousel('next')`—This cycles to the next item.

Carousel Events

Bootstrap exposes two events when you use the Carousel plugin:

- ► `slide.bs.carousel`—This fires immediately when the `slide` instance is invoked.

- ► `slid.bs.carousel`—This fires when the carousel has completed the slide operation.

Both events have two additional properties:

- ► `direction`—The direction the carousel is sliding, either `left` or `right`

- ► `relatedTarget`—The DOM element that is being slid into place as the active element

Carousels on the Web

Before you head off to create a carousel for every web page you own, you should know that there are drawbacks to using carousels. Customers don't always understand them, they can be difficult to use, and sometimes they even drive people away from your website.

Using Bootstrap does not prevent these problems. In fact, some would argue that the default Bootstrap carousel behavior is hard to use and annoying.

Carousel Best Practices

There are several best practices you can put into place to ensure that your carousels work as well as possible and don't annoy your readers. Here they are:

▶ **Keep the items in your carousel the same**—If you're doing a showcase of sale items, then cycle through them. Don't switch to an ad for your latest email newsletter followed by a list of blog posts. Your customers are more likely to take notice of a carousel that is cycling through a group of similar items than one that is random.

▶ **Your customers don't read as fast as you do**—Even though they might read fast, they don't have the intimate knowledge of the content that you do. The most common problem with carousels on web pages is that they cycle too quickly. A good rule of thumb is to pick a time that feels slow to you and then double that time. Yes, it will feel glacially slow to you as the designer, but your customers might actually stop and look at the content.

▶ **But a better solution is don't turn on automatic sliding at all**—Most people hate auto-play sliders. In fact, in a study by the Nielsen Norman Group (http://www.nngroup.com/articles/auto-forwarding/), they discovered that carousels that autoplayed were much more likely to be ignored by customers because they were too hard to read. The best carousels should be off by default and allow the customer to choose when to cycle to the next slide.

▶ **If you must have autoplay carousels, make it easy to pause them**—With the Bootstrap carousels, this means you should *not* turn off keyboard controls or the pause function. It also helps to have explanatory text that tells customers how to stop the carousel. Don't assume people will just know to mouse over the top of it.

▶ **Make the navigation obvious**—The standard navigation elements in a Bootstrap carousel include the indicator dots and the right and left arrows. These are good, but if the slides are too light, they can be hard to see. Just the indicator dots is *not enough* to be clear to most people that there is more content. The best navigation uses actual words like "Previous" and "Next" along with arrows or icons to make it clear that more content is available.

▶ **Don't go overboard on the number of slides**—It can be tempting to add as many slides as you can to get as much information on the page as possible. But remember that just because it's in the HTML doesn't mean that a customer sees it. If your average customer stays on your page for 30 seconds and you have a carousel cycling between slides every 6 seconds, your customer will be shown only 5 slides in that time. Whether you have 6 slides or 600 won't matter because most of your customers will never see them. But if you have 600 slides on the page, customers will have to download all 600, and that will affect their willingness to stay on the page.

Be sure your site really needs a carousel. It can be argued that no website needs a carousel, but don't add one just because they are cool or to appease three different groups in your company.

As you'll see in the following section, carousels don't solve the problems most people think they do—and they can cause more problems.

Problems with Carousels and Some Solutions

The most common reason people add carousels to a home page is to get a diverse set of content "visible" on the page without taking up a lot of screen real estate. But the reality is, even with autorotating carousels, readers ignore them in favor of content elsewhere on the page.

In one study only 1% of readers interacted with the carousel at the top of the page. Of that 1%, 89% of them interacted with only the first slide (http://erikrunyon.com/2013/07/carousel-interaction-stats/). When you consider that carousels are often huge, taking up one-third or more of the visible space on many home pages, it is very disappointing to see a 1% interaction rate. In effect, the carousel is forcing readers to scroll down to find content they want to look at. We all know that if a reader doesn't find what she wants, she will leave.

Carousels are also an accessibility nightmare. Screen readers find them hard to read, and they are often very difficult to navigate with the keyboard. Even the Bootstrap documentation recommends using other methods if you need to be accessible.

"The carousel component is generally not compliant with accessibility standards. If you need to be compliant, please consider other options for presenting your content." (http://getbootstrap.com/javascript/#accessibility-issue)

 If you have a lot of content you want to display on your home page, there are a few other options you can use besides a carousel. Some popular solutions include

- ▶ **A grid of content blocks**—Use your Bootstrap grid to create small static boxes of varying sizes that take up the same space the carousel would have. You can make the blocks all the same size or create a larger box to highlight one element more.

- ▶ **A primary banner**—By focusing on one primary message, your website will be clearer. If you need to get the other information on the page, though, you can add smaller thumbnail images of the other content for customers to click on to learn more.

- ▶ **An overflow pattern**—Set up a design where everything in the carousel gets the same amount of space allocated to it. The content that doesn't fit is placed behind a link that says very clearly "more content" or "read more" or an arrow indicating that.

- ▶ **Separate landing pages**—From a search engine optimization standpoint, this is your best solution. Rather than trying to target multiple audiences with one page, create separate pages for each audience.

Finally, you should test your carousels to make sure they are working for your site. Your CEO might be happy that his pet project is listed on the home page, but what's he going to say when no one is visiting that project even with the home page boost? Read the analytics on your site and confirm whether your readers interact with the carousels or ignore them. After you have the data, you'll be able to more effectively argue for a replacement.

Summary

In this hour you learned how to create a banner of rotating content on your website using the Bootstrap Carousel plugin. You also learned how to create basic image slideshows both with and without captions as well as fancier HTML slides with images and text. You learned how to add multiple carousels to a page and which options (see Table 20.1), methods (see Table 20.2), and events (see Table 20.3) are available for your scripts. You also learned about several new classes that help build your carousels (see Table 20.4).

TABLE 20.1 Carousel Options

Option	Description
interval	The amount of time, in milliseconds, to delay between cycling slides. The default is 5000; if you use false, the slides won't cycle at all.
keyboard	A Boolean option that determines whether the carousel should react to keyboard commands. The default is true.
pause	A string to define when the carousel should pause. The default is hover and will make the carousel pause when the mouse is over the slides.
wrap	A Boolean option that determines whether the carousel should wrap to the beginning after cycling through the last item. The default is true.

TABLE 20.2 Carousel Methods

Method	Description
$().carousel(options)	Activates the carousel and begins cycling through the slides
.carousel('cycle')	Cycles through the carousel from left to right
.carousel('next')	Cycles to the next item
.carousel(number)	Cycles to the specific item number with zero (0) being the first
.carousel('pause')	Stops the cycling through items
.carousel('prev')	Cycles to the previous item

TABLE 20.3 Carousel Events

Event	Description
slide.bs.carousel	Fires immediately as the slide instance is invoked
slid.bs.carousel	Fires after the carousel has completed the slide transition

TABLE 20.4 CSS Classes

Class	Description
.carousel	Defines the element as a carousel.
.carousel-caption	Defines the element as a caption for a carousel item.
.carousel-control	Defines a carousel control element. Used with the .right and .left classes to define specific controls.
.carousel-indicators	Defines the carousel indicator icons used to subtly identify how many slides are available and navigate between them.
.carousel-inner	Identifies the group of slides in a carousel.
.item	Indicates the element is a slide in a carousel.
.slide	Tells the browser to add the slide transition on the carousel.

Workshop

The workshop contains quiz questions to help you process what you've learned in this hour. Try to answer all the questions before you read the answers.

Q&A

Q. I built a carousel from your code Listings 20.1, 20.2, and 20.3, and it doesn't look like the carousel you showed in Figure 20.1. Why is that?

A. I added a few custom styles to the basic carousel to make the slides display nicely. Listing 20.7 shows my custom styles.

LISTING 20.7 Custom Styles for a Basic Carousel

```
.item h1 {
  width: 10em;
  margin-left: auto;
  margin-right: auto;
  margin-bottom: 2em;
}
.carousel-indicators li, .carousel-indicators .active {
  border-color: #000;
}
.carousel-indicators .active {
  background-color: #000;
}
```

If you add these styles to your custom style sheet, your carousel should look just like mine.

Q. **The carousel indicators are always at the bottom of the slides, but what if I want to place them at the top?**

A. You can customize their placement by adjusting your CSS. For example, if you want to place them at the top of the slide, you could add the following line to your CSS file:

```
.carousel-indicators { top: 10px; }
```

Quiz

1. Which of the following is required to create a carousel in Bootstrap?

    a. An element with the `.carousel` class on it

    b. An element with the `.carousel.slide` classes on it

    c. A unique ID on the element with the `.carousel` class on it

    d. A unique ID of the element with the `.carousel.slide` classes on it

2. To create a caption, you place the `<figcaption class="carousel-caption">` element where?

    a. Inside the `.carousel` element.

    b. Inside the `.carousel-inner` element.

    c. Inside the `.item` element.

    d. Nowhere; you can't use a `<figcaption>` element for a caption.

3. Which of these classes creates a group of small dots at the bottom of the carousel?

    a. `.carousel-caption`

    b. `.carousel-control`

    c. `.carousel-indicators`

    d. `.carousel-inner`

4. What does the `data-slide-to` attribute do?

    a. It points to the number of the slide to cycle to.

    b. It indicates the number of the current slide.

    c. It indicates how many times the carousel should cycle.

    d. It points to the slide controls.

5. True or False: Controls and indicators are required on carousels.

    a. True, because both are required.

    b. False, because only controls are required.

    c. False, because only indicators are required.

    d. False, because neither are required.

6. How do you add multiple carousels to a single page?

   a. Call each carousel with a separate script.

   b. Give each carousel a unique ID.

   c. Add a new class to each new carousel.

   d. You can't. Only one carousel is allowed per page.

7. Which is a better way to implement a carousel: with an interval of 10,000 milliseconds or with an interval of `false`?

   a. They do the same thing.

   b. A 10,000-millisecond pause is best because it gives customers a long time to read the slide.

   c. An interval of `false` is best because it lets the customers decide when to move to the next slide.

   d. Neither; best practice says to apply both.

8. Which of the following provides the best navigation options?

   a. Just carousel indicators

   b. Carousel indicators with iconic controls

   c. Carousel indicators with text controls

   d. Carousel indicators with text controls and written instructions

9. What is the biggest problem with carousels?

   a. Customers ignore them.

   b. The content doesn't display.

   c. The animations are distracting.

   d. They don't work on older browsers.

10. True or False: Carousels are accessible.

Quiz Answers

1. c. A unique ID on the element with the `.carousel` class on it. The `.slide` class is not required, but the unique ID is.

2. c. Inside the `.item` element.

3. c. `.carousel-indicators`

4. a. It points to the number of the slide to cycle to.

5. d. False, because neither are required.

6. b. Give each carousel a unique ID.

7. c. An interval of `false` is best because it lets the customers decide when to move to the next slide.

8. d. Carousel indicators with text controls and written instructions.

9. a. Customers ignore them.

10. False.

Exercises

1. Add a carousel to your website. Be sure to include indicators and controls. Evaluate the slides and include only the three most critical ones. Turn off the autoplay feature, so that your customers can decide when to cycle the slides.

2. Add tracking to your analytics to evaluate how your carousel does with your customers. After a month, check to see how many times the carousel was clicked on and which slides were the most popular. If you want, change the carousel to autoplay and see whether that changes the analytics.

Customizing Bootstrap and Your Bootstrap Website

What You'll Learn in This Hour:

▶ How to use your own CSS to customize Bootstrap sites

▶ How to customize the Less files and jQuery plugins your site uses

▶ How to customize the CSS with the Bootstrap Less variables

▶ How to download your customizations with the Bootstrap Customizer

Bootstrap is a very flexible framework that you can use to create responsive websites. It comes with a standard design and color scheme that many people like. But if you want to create a site that is more customized, there are several ways to do it. You can write your own CSS and apply that after the Bootstrap CSS. Or you can use the Bootstrap customization page to create a custom build of Bootstrap that has the styles and colors you need for your website.

In this hour, you learn how to add your own custom CSS files to Bootstrap web pages. You also learn how to go through the entire Bootstrap customization form to create a custom build for your website.

Using Your Own CSS

The easiest way to customize Bootstrap is to use your own CSS. To do this, download the standard Bootstrap files (refer to Hour 2, "Downloading and Installing Bootstrap") and then add your own CSS to the pages.

You can add your own custom CSS in three ways:

▶ In the HTML elements using the `style` attribute

▶ In the `<head>` of your pages using the `<style>` element

▶ In a linked style sheet

Using the `style` attribute is the easiest way. You simply place the styles you want as a value for the attribute. For instance, if you want to change a paragraph color to red, you write

```
<p style="color: red;">
```

But if you use the `style` attribute, you have to add it to every element you want to change. A better method is to use a `<style>` element in the head of your document. You can use this to style every paragraph on the page or just the ones with a specific class or ID, or you can use it to set up more complex CSS selectors.

NOTE

Learning CSS

If you don't know CSS, there are a few things you can do. You should focus on understanding selectors. For instance, a class selector starts with a period (.) and an ID selector starts with a hash (#). Don't worry as much about memorizing the style properties. It's easier to find a site you like that lists CSS style properties and reference that when you need one. A good book to get started with CSS is *Sams Teach Yourself HTML and CSS in 24 Hours* by Julie C. Meloni.

To style a paragraph to have red text, you must add the class `.red` to it (`<p class="red">`) and then add the following CSS to the head of your document, below the Bootstrap CSS link:

```
<style>
  .red { color: red; }
</style>
```

But the best way to add CSS to your web pages is with an external style sheet that is linked in the `<head>` of the document:

```
<link href="css/styles.css" rel="stylesheet">
```

Just make sure that the `href` points to your style sheet file. Then, insert your styles just like in the `<style>` element:

```
.red { color: red; }
```

The most effective way to use your own style sheets with a Bootstrap site is to style the same classes that Bootstrap does. Don't forget to use media queries to affect specific device sizes. Figure 21.1 shows a Bootstrap page with no additional styles. As you can see, right out of the box, Bootstrap does a lot to help your pages look great.

FIGURE 21.1
A Bootstrap page with no extra styles.

The challenge is that with no changes, every Bootstrap site looks the same—and that can get a bit bland. Figure 21.2 shows the same site as in Figure 21.1 but with a few colors changed with CSS.

FIGURE 21.2
A Bootstrap page with extra styles.

Listing 21.1 shows the CSS I used to create Figure 21.1. This is the most basic of style sheets and would not be sufficient to style an entire Bootstrap website, but it works great for a start.

LISTING 21.1   CSS to Style the Dandelion Site

```css
body {
  background-color: #4C521D;
}
body > .container {
  background-color: #fff;
}
.jumbotron {
  background: #6c702d;
  background: -moz-linear-gradient(top, #6c702d 0%, #dfdfdf 100%);
  background: -webkit-gradient(linear, left top, left bottom,
    color-stop(0%,#6c702d), color-stop(100%,#dfdfdf));
  background: -webkit-linear-gradient(top,
    #6c702d 0%,#dfdfdf 100%);
  background: -o-linear-gradient(top, #6c702d 0%,#dfdfdf 100%);
  background: -ms-linear-gradient(top, #6c702d 0%,#dfdfdf 100%);
  background: linear-gradient(to bottom, #6c702d 0%,#dfdfdf 100%);
  filter: progid:DXImageTransform.Microsoft.gradient(
    startColorstr='#6c702d', endColorstr='#dfdfdf',
    GradientType=0 );
  color: #000;
}
.jumbotron h1 {
  color: #D8AA10;
  text-shadow: 4px 4px 4px #000;
}
.quote {
  background-color: #6C702D;
  color: #ddd;
}
.carousel-control.left {
  background-image: -webkit-linear-gradient(left,
    rgba(170,140,23,1) 0,rgba(0,0,0,.0001) 100%);
  background-image: -o-linear-gradient(left,rgba(170,140,23,1) 0,
    rgba(0,0,0,.0001) 100%);
  background-image: -webkit-gradient(linear,left top,right top,
    from(rgba(170,140,23,1)),to(rgba(0,0,0,.0001)));
  background-image: linear-gradient(to right,rgba(170,140,23,1) 0,
    rgba(0,0,0,.0001) 100%);
  filter: progid:DXImageTransform.Microsoft.gradient(
    startColorstr='#80000000', endColorstr='#00000000',
    GradientType=1);
}
```

```
.carousel-control.right {
  background-image: -webkit-linear-gradient(left,
    rgba(0,0,0,.0001) 0,rgba(170,140,23,1) 100%);
  background-image: -o-linear-gradient(left,rgba(0,0,0,.0001) 0,
    rgba(170,140,23,1) 100%);
  background-image: -webkit-gradient(linear,left top,right top,
    from(rgba(0,0,0,.0001)),to(rgba(170,140,23,1)));
  background-image: linear-gradient(to right,rgba(0,0,0,.0001) 0,
    rgba(170,140,23,1) 100%);
  filter: progid:DXImageTransform.Microsoft.gradient(
    startColorstr='#00000000', endColorstr='#80000000',
    GradientType=1);
}
h1, h2, h3, h4, h5, h6, a, a:link {
  color: #D8AA10;
}
.panel-default>.panel-heading {
  background-color: rgba(179,196,170,.25);
}
.btn.btn-info {
  background-color: #d8aa00;
  border-color: #aa8c10;
}
```

▼ TRY IT YOURSELF

Changing Bootstrap Styles

Sometimes the best way to change the Bootstrap styles is to use the same selectors as Bootstrap does. But it can be difficult to figure out what those selectors are or what they do. In this Try It Yourself, you learn how to use your web browser to discover the selectors Bootstrap uses and then how to adjust your CSS to override the Bootstrap CSS:

1. Build your web page using Bootstrap, and then open it in Chrome or Safari.

2. Right-click the web page element you want to style, and select Inspect Element, as in Figure 21.3.

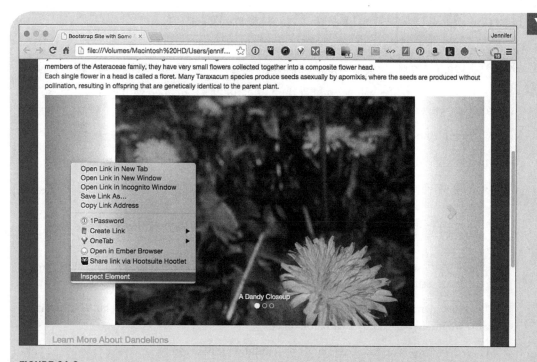

members of the Asteraceae family, they have very small flowers collected together into a composite flower head. Each single flower in a head is called a floret. Many Taraxacum species produce seeds asexually by apomixis, where the seeds are produced without pollination, resulting in offspring that are genetically identical to the parent plant.

Open Link in New Tab
Open Link in New Window
Open Link in Incognito Window
Save Link As...
Copy Link Address

1Password
Create Link
OneTab
Open in Ember Browser
Share link via Hootsuite Hootlet

Inspect Element

A Dandy Closeup

Learn More About Dandelions

FIGURE 21.3
Inspecting an element in Chrome.

3. Click the element in the HTML column (on the left); you will see the CSS that styles it on the right.

4. Copy the applicable CSS into your style sheet.

5. Change the styles to reflect your design.

As long as your CSS is loaded after the Bootstrap CSS, your styles will override the Bootstrap styles.

Using CSS to customize your Bootstrap site is a quick and easy way to do it. Although you can use the `style` attribute, the `<style>` tag, or external style sheets to get the styles onto the page, the best method is to use external style sheets. This helps your pages to load more quickly, and you can share more styles across all the pages of your site.

Using the Bootstrap Customizer

Your other option to customize your Bootstrap website is to customize your Bootstrap files themselves. This can take more time than just updating a few CSS styles, but it has a few advantages:

▶ Custom Bootstrap files will be smaller than original files combined with custom CSS, so your site will load more quickly.

▶ By customizing the JavaScript, you can reduce the size even more.

▶ The CSS will be more comprehensive, making future changes more likely to just work.

▶ You can change features that aren't localized so that your site is more consistent.

But as with everything, there are also some drawbacks to using the Bootstrap customization script:

▶ Because you have to use the generator, it can be a different workflow and thus slow down your design schedule.

▶ The online generator does not have any way to preview your styles, so it can be difficult to know whether your changes will look good.

▶ When Bootstrap is updated, you have to rebuild your files with your styles.

To customize the Bootstrap files, you should use the generator from Bootstrap (http://getbootstrap.com/customize/). The following sections describe how to use the customizer.

CAUTION

Customizer Doesn't Work in Safari

If you browse with Safari, you will need to get a different browser to use the customizer. In Safari saving the generated zip file doesn't work. The script then opens the zip in a second tab as a data URI that has to be manually saved. Because this is very confusing and difficult, the Bootstrap team decided to disable Safari. If you normally use Safari, you need to switch to Chrome, Firefox, or Opera to customize Bootstrap.

Less Files and jQuery Plugins

The first part of the customizer, as shown in Figure 21.4, is the Less files you want to include in your website.

FIGURE 21.4
Get Bootstrap Customizer—Less Files section.

This section customizes the Less files for the CSS, components, and JavaScripts of Bootstrap. These were all covered in the previous hours of this book.

To create the most efficient set of Bootstrap files, you should deselect all the items you don't use and don't plan on using on your website. Remember that you can always recompile your customizations if you want to add something back later.

The jQuery plugins section (pictured in Figure 21.5) enables you to customize the JavaScript for your site. If you removed a component or JavaScript from the Less files, you should remove them from the JavaScript as well.

FIGURE 21.5
Get Bootstrap Customizer—jQuery plugins section.

The customizer produces two JavaScript files: `bootstrap.js` and `bootstrap.min.js`. It's best to use the minified file (`bootstrap.min.js`) in production because it's smaller and more efficient.

Remember that even the customized JavaScript requires jQuery installed on the site as well.

Less Variables

This is the part of the customization form that gets interesting. As you can see from the navigation in Figure 21.6, there are more than 30 types of things you can customize.

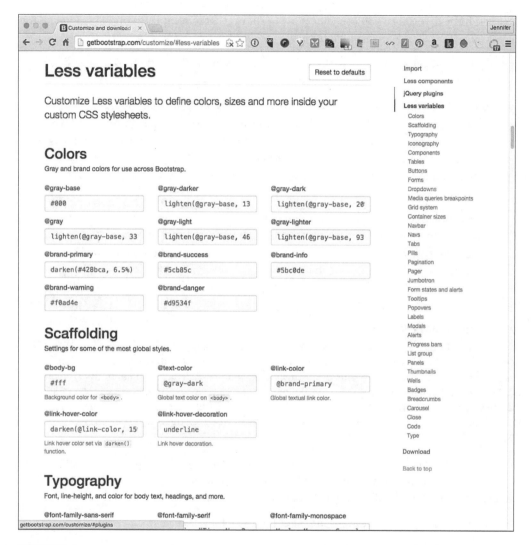

FIGURE 21.6
Get Bootstrap Customizer—Less variables.

The first thing you should do is come up with your color scheme. Here are two free online tools that let you create a color scheme from a color or from a photo:

▶ **Paletton (http://paletton.com/)**—Paletton lets you pick a color scheme with a starting color and a configuration (such as monochromatic or complementary). This is a great generator for when you have a color you'd like to focus your palette on.

▶ **PaletteGenerator (http://palettegenerator.com/)**—PaletteGenerator lets you upload an image, select a section of it, and generate a palette of 2–10 colors.

After you have your colors, you can start playing with the customizer Less variables. Some of the colors you will need to define with your palette include

▶ **Contextual colors**—These are the `@brand-*` colors in the first section. You must define a color for `primary`, `success`, `info`, `warning`, and `danger`.

▶ **Background, text, and link colors**—These are set in the Scaffolding section. The defaults for these values are white (`#fff`), dark gray using the `@gray-dark` variable, and whatever you set for `@brand-primary`. You also can change the link hover color, but leaving it `darken(@link-color, 15%)` uses a Less mixin to use the same link color only 15% darker. You will learn more about Less mixins in Hour 23, "Using Less and Sass with Bootstrap."

▶ **All the other components**—This includes tables, buttons, forms, dropdowns, navbars, navs, tabs, pills, pagination, pagers, the Jumbotron, form states and alerts, tooltips, popovers, labels, modals, alerts, progress bars, list groups, panels, thumbnails, wells, badges, breadcrumbs, carousels, close, code, and type.

NOTE

Keep Your Colors Consistent

One good rule of thumb is to use the same colors for similar things. The Bootstrap Less variables make that easier because you need to define a variable only once and then you can use it in all the other similar variables. For example, if you want all your `@*-primary` elements to be colored `#D1A40C` (a dark mustard yellow), you can define `@brand-primary` in the "Colors" section and then use `@brand-primary` instead of a color code in the rest of the customization form. I don't know about you, but it's much easier for me to remember `@brand-primary` than `#D1A40C`.

Another way that you can customize your designs so they don't look like a typical Bootstrap site is to adjust the typography and fonts. The "Typography" and "Iconography" sections are the first place to start. You should come up with three font stacks:

▶ `@font-family-sans-serif`—Sans-serif fonts like Helvetica Neue and Arial

▶ `@font-family-serif`—Serif fonts like Times New Roman and Georgia

▶ `@font-family-monospace`—Monospace fonts like Courier New and Monaco

Bootstrap uses the sans-serif font stack as the default font family. But you can change this to either the serif or the monospace fonts by changing the variable in the `@font-family-base`.

Another thing you'll notice in the "Typography" section are the terms `ceil` and `floor`. These refer to rounding numbers up or down, respectively. For example, `@font-size-large` is defined as `ceil((@font-size-base * 1.25))`. This means that the large font size is the base font size (`@font-size-base`) times 1.25 and then rounded up (`ceil`). The `floor` term means the number should be rounded down.

Finally, you can look at the layout designs. You can use variables to adjust the media query breakpoints, the grid system, and container sizes. These let you customize your layout including how many columns in your grid, the different widths for media queries, and the size of your containers. I recommend not changing the @screen-*-max variables. They are currently set to one pixel smaller than the next highest size. In other words, the maximum for extra small screens is the minimum for the small screens minus one. This ensures that there aren't any gaps in the breakpoints that wouldn't be styled.

Downloading and Installing Your Customizations

After you've filled out all your customizations, you download the files by clicking the Compile and Download button shown in Figure 21.7.

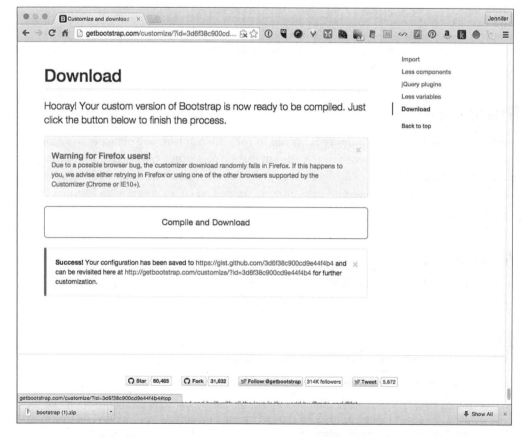

FIGURE 21.7
Get Bootstrap Customizer—download and compile.

After your configuration has been compiled, a zip file is downloaded to your computer and you're given a URL where you can go back and continue to edit that custom build as necessary.

When you open the zip file, you'll find a `config.json` file you can use to upload your configuration later. You'll also have custom CSS, fonts, and JavaScript files. Install those on your website just like you did in Hour 2.

The rest of the files in the zip file look exactly like the file listing you get when you install a standard version of Bootstrap. You will get both full and minified versions of the JavaScript and CSS as well as multiple copies of the icon fonts. Best practices suggest that you use the minified versions of the files in production to keep your download speed as fast as possible.

Using a Third-Party Bootstrap Customizer

Many people find the default customizer on the Bootstrap website to be annoying because it doesn't offer any way to preview your theme choices before you download the files.

Luckily, a lot of customizers are available on the Web that can help you see your styles before you load them on your website. I go into more detail on several Bootstrap customizers in Hour 24, "Going Further with Bootstrap."

Summary

This hour explained how to customize Bootstrap so your site can look and act more like a unique website and less like every other Bootstrap site out there. You learned how to add your own CSS styles in the document in `style` attributes, `<style>` elements, and external style sheets. Then you learned how to use the Bootstrap customizer to create and compile your own custom version of Bootstrap.

Workshop

The workshop contains quiz questions to help you process what you've learned in this hour. Try to answer all the questions before you read the answers.

Q&A

Q. When I added my custom CSS, it didn't override the Bootstrap styles. What can I do?

A. The most common reasons custom CSS doesn't override the Bootstrap ones are

- ▶ **The custom CSS is loaded before the Bootstrap CSS**—Your CSS style sheet must be the last style sheet loaded so that its styles override any others. You can ensure this by putting the style sheet right before the closing `</head>` tag in your document.

- ▶ **The Bootstrap styles are more specific than your styles**—For example, say you had a Jumbotron with an `<h1>` element with a `<small>` element inside it. You could change the color of the `<h1>` by simply assigning a text color to the Jumbotron with

`.jumbotron { color: #D8AA10; }`. But the `<small>` text would not be the correct color, as shown in Figure 21.8, because Bootstrap has a more specific style, `h1 small { color: #777; }`. It is more specific because it styles all `<small>` elements that are inside an `<h1>` not just any text in a Jumbotron.

FIGURE 21.8
Custom CSS fails to style everything.

You can learn more about CSS specificity on the W3C website http://www.w3.org/TR/css3-selectors/#specificity.

Q. Why would I want to use my own custom CSS? It seems like there are a lot of drawbacks to it.

A. Yes, custom CSS can be difficult to add and not cover every situation, but it is often a lot faster to use. Custom CSS is often easier to include in how you normally build web pages, and if you come across a problem like the font color in Figure 21.8, you can quickly fix it. You don't need to know Less or understand the variables. I generally use custom CSS for my design phase, and after the entire site is designed, I work on a custom build of the Bootstrap files.

Quiz

1. Is this HTML a good way to override the Bootstrap CSS with your own?

```
<link href="styles.css" rel="stylesheet">
<link href="css/bootstrap.min.css" rel="stylesheet">
```

    a. Yes, this would work perfectly.

    b. No, both lines are missing a required attribute.

    c. No, they are in the wrong order; the Bootstrap file should be first.

    d. No, the custom style sheet must be in the same directory as the Bootstrap file.

2. Which of the following style properties are not allowed with Bootstrap?

    a. CSS speech properties

    b. All CSS3 properties

    c. All of the above

    d. None of the above

3. Which of these is the best way to override Bootstrap style properties?

    a. Inline styles using the `style` attribute.

    b. In page styles using the `<style>` property.

    c. External style sheets.

    d. None; they are all equally good.

4. Which of these is not a benefit to using the Bootstrap customizer?

    a. The CSS is automatically updated.

    b. The file size is smaller.

    c. You can load only the plugins you need.

    d. You can be sure to adjust every feature, even ones you aren't using right away.

5. Which is a reason you wouldn't want to use the customizer?

    a. It doesn't work on Macintosh computers.

    b. It doesn't include a preview to check your styles.

    c. It forces you to use some components and plugins you might otherwise remove.

    d. It customizes only some of Bootstrap.

6. Why would you deselect one of the Less files?

    a. It includes a plugin you don't need.

    b. It includes a plugin that breaks your site.

    c. It includes a component you don't need.

    d. It includes a component that breaks your site.

7. True or False: You can compile Bootstrap to include jQuery with the customizer.

8. Which of these is the name of the file created by the customizer that would include the Bootstrap Tooltip plugin on the site?

    a. `bootstrap.css`

    b. `bootstrap.js`

    c. `tooltip.css`

    d. `tooltip.js`

9. True or False: You must know Less to use the Bootstrap customizer.

10. What do the Less variables configure?

    a. The jQuery plugins

    b. How the Bootstrap minifies the JavaScript

    c. Only the website colors

    d. All the Bootstrap CSS

Quiz Answers

1. c. No, they are in the wrong order; the Bootstrap file should be first.

2. d. None. You can use all CSS properties with Bootstrap.

3. c. External style sheets.

4. a. The CSS is automatically updated.

5. b. It doesn't include a preview to check your styles.

6. c. It includes a component you don't need.

7. False.

8. b. `bootstrap.js`. The customizer creates a `bootstrap.js` file that includes the `tooltip.js` information.

9. False. Although it might help, most of the variables are fairly self-explanatory.

10. d. All the Bootstrap CSS.

Exercises

1. Add some custom styles to your Bootstrap site using an external style sheet or `<style>` tag.

2. Make some customizations with the Bootstrap customizer, and download and install them on your site. Remember that if you don't like them, you can always revert to the original by downloading the full Bootstrap files.

Making Bootstrap Accessible

What You'll Learn in This Hour:

- ▶ What web accessibility is
- ▶ General accessibility rules
- ▶ How to improve Bootstrap accessibility
- ▶ Specific things you can do to make your Bootstrap site accessible

Accessibility is one of those web design tasks that many designers either don't know about or ignore, hoping it will solve itself. To some extent, because you're using Bootstrap, your site will be more accessible than not. But there are still some things to do to ensure that people using assistive technology (AT) can use your site as well as others.

In this hour you learn what accessibility means when it comes to websites and how to make specific Bootstrap components and plugins accessible. You also learn how to evaluate your site for accessibility so you know what you do and don't need to do.

What Is Accessibility?

When designing web pages, you should think *inclusion*. Including as many people as possible means that you will have the largest possible audience. This means taking into consideration things like age, economic situation, location, and language. It also means considering people with disabilities. Accessibility usually focuses on people with disabilities, but, in general, web designers should focus on inclusive design.

According to Johns Hopkins University, "Web accessibility refers to a standard of inclusive website development based on the idea that information should be equally available to all people, regardless of physical or developmental abilities or impairments" (http://webaccessibility.jhu.edu/what-is-accessibility/).

Accessible Design Practices

When you practice inclusive or accessible design, you need to keep a few basic things in mind:

▶ **How do your pages look?** Ask yourself whether there is enough contrast between colors. Are the colors different enough to be clear? How would a color-blind person perceive the page?

▶ **How usable are your pages?** Consider whether your pages require a mouse, or can a keyboard navigate the links? Are the elements far enough apart to be clicked or tapped on for someone with reduced motor skills or even "fat fingers"?

▶ **How does your website sound?** Think about whether sound is required to use your site. If someone used a screen reader, would all the content be available to them?

▶ **How complicated is your website?** Consider whether a lot of steps are required to complete various tasks. Are there things that might be distracting or make it difficult to focus on the page?

The nice thing about accessibility is that you don't have to make a lot of major changes to your website. You should just be aware of how your site might look or sound to someone with differing abilities to your own.

Here are the guidelines for building an accessible web page:

▶ **Provide equivalent alternatives for auditory and visual content**—If you have video or sound content on your site, you should have a text version for screen readers to read. For example, if you have a video, it should have captions and you should write out the text.

▶ **Do not rely on color alone to convey information**—You should also include text that provides the same information. For example, if you mark warning messages with a red background color, you should also include the word *Warning* on the page.

▶ **Use HTML, CSS, and JavaScript correctly, as per the specifications**—Invalid code can cause problems for screen readers that might not be as current on browser trends. For example, even though you can write an HTML table without the closing `</table>` tag and have it render on most browsers, you should include the closing tag.

▶ **Clarify natural language usage**—Be sure to identify the language of your web page with the `lang` attribute. For example, on an English language page, you should write `<html lang="en">`.

▶ **Make sure your tables are responsive**—It's best to use tables for tabular data rather than layout.

▶ **Provide alternatives for new technologies**—Fallback options and alternative text are important.

▶ **Allow users to control time-related content changes**—Blinking, rotating, and animated text and images can be difficult to read.

▶ **Design for device independence**—Responsive web design accounts for a lot of this, but this also includes things like text links as an alternative to image maps and using the `tabindex` attribute to help people without mice tab between links.

▶ **Provide contextual information, even if you think it won't be visible**—Things like titles for tables and links, descriptive text, and alternative text all give assistive devices more information for your customers.

▶ **Use clear and consistent navigation structures**—This includes links in the text of your pages. Remember that screen readers might read links as navigation and remove them from the context of surrounding text, so links that say only "click here" do not provide enough information.

▶ **Keep your pages as simple and clear as possible**—This will ensure that they are easy for everyone to use.

The WAI-ARIA and Accessibility

WAI-ARIA is the Web Accessibility Initiative's Accessible Rich Internet Applications Suite. It defines a way to make the Web more accessible to people with disabilities. This is done through the use of a number of attributes, and most ARIA attributes start with `aria-*`. Here is a list of the attributes most often used to create accessible pages:

▶ `aria-controls`—This attribute tells assistive technology (AT) that the element is a control element. It contains the ID of the controlled element.

▶ `aria-describedby`—This attribute tells AT where the descriptive text is. It contains the ID of an element that holds the descriptive text. For example, if you have a table and provide descriptive text in a `<div id="tableDesc">`, you point to it in the table tag: `<table class="table" aria-describedby="tableDesc">`.

▶ `aria-expanded`—This attribute tells AT that the element is an expandable element like an accordion. Set it to `true` if the element is open or `false` if it's closed.

▶ `aria-hidden`—The `aria-hidden="true"` attribute tells the AT device that the element is hidden.

▶ `aria-invalid`—If you have a form control that is invalid, you can assign the `aria-invalid="true"` attribute to let AT know that it is invalid.

▶ `aria-label`—The `aria-label` attribute points to the ID of the element it is labeling. It is similar to the `for` attribute on `<label>` tags.

▶ `aria-labeledby`—The `aria-labeledby` attribute points to the ID of the element by which it is labeled. It is similar to the `aria-describedby` attribute.

▶ `aria-pressed`—The `aria-pressed="true"` attribute tells AT that a button with the `.active` state is pressed or active.

▶ `role`—The `role` attribute tells the assistive device what the element's purpose is. This is used to ensure that screen readers and other AT reference those elements correctly. It will not change how an element looks to non-AT browsers. For example, `<h1 role="button">I am a Button</h1>` will display as an `<h1>` headline, but AT will see it as a button.

▶ `title`—The `title` attribute is an HTML attribute you can use to give any element a title. Some non-AT browsers will display the title on some elements, so be sure to test this.

There are many more ARIA attributes you can use to make your pages accessible.

Accessible Design in Bootstrap

Here are a number of things you can do with your Bootstrap web pages to start making your pages more accessible.

▶ Avoid using pop-up windows as much as possible. Bootstrap does not provide any built-in plugins for pop-up windows, and the plugins it does provide keep accessibility in mind.

▶ Use form labels and group form elements with the `<fieldset>` tag. It's best if your labels come before the form fields. Bootstrap forms work best with the `<label>` tag, as you learned in Hour 9, "Styling Forms."

▶ Use placeholders in form input tags such as `<textarea>` and `<input>` until all user agents handle empty form controls correctly.

▶ Use alternate text on all images, video, audio, and other plugins.

▶ Provide text alternatives for tables.

▶ Include non-linked space or text characters between adjacent links.

Skip Navigation

Skip links or skip navigation is an important part of an accessible web page design, especially in designs that contain a large group of navigation links at the top of the page. A *skip link* lets someone using AT skip the navigation and get directly to the content of the page. Bootstrap lets you easily create a skip link that is shown to only screen readers and other AT. Listing 22.1 shows how you would add a skip link.

LISTING 22.1   A Bootstrap Skip Link

```
<body>
  <a href="#main" class="sr-only sr-only-focusable">Skip to primary content</a>
  ...
  <div class="container" id="main" tabindex="-1">
    <!-- The main page content -->
  </div>
</body>
```

As is shown in Listing 22.1, the first thing inside the `<body>` tag is the skip link. This points to an anchor lower on the page. The classes `.sr-only` and `.sr-only-focusable` hide the link from browsers that are not screen readers (as you learned in Hour 13, "Bootstrap Utilities"). Then the first container is given the `id` of main, and that is where the link points.

CAUTION

Don't Forget the `tabindex` Attribute

A few browsers (Chrome and Internet Explorer) will not focus on the skip link programmatically without the `tabindex` attribute. Leaving off the `tabindex` makes any in-page link you are using inaccessible for keyboard users, not just the skip navigation. So it's a good idea to include it on any element where the `id` will be used as a link target. Bootstrap also recommends suppressing the visible focus indication on the target of your skip link with the CSS: `#main:focus { outline: none; }`.

Nested Headings

Even though HTML5 suggests that you can use the `<h1>` tag for every headline on a page using CSS to provide styles, it is better to use the `<h#>` tags in a hierarchy. The `<h1>` tag is first and should be the title for the entire page. Then use `<h2>` tags to title the page sections, with `<h3>` as subtitles and so on. If you keep these tags in a logical order, then screen readers can create a table of contents out of them.

Color Contrast

Many designers think of color blindness when they are designing their color scheme, but contrast can sometimes fall by the wayside. Some of the default Bootstrap colors have a low contrast ratio. The recommended minimum ratio is 4.5:1.

You should check the colors on the following Bootstrap elements to ensure that they have a good contrast ratio:

▶ Highlighting colors for basic code blocks (refer to Hour 7, "Bootstrap Typography")

▶ The `.bg-primary` contextual background helper class (refer to Hour 7)

▶ The default link color on a white background

NOTE

Bootstrap Is Always Changing

During the writing of this book, the Bootstrap customizer was changed to make the @brand-primary colors have better contrast. Bootstrap is constantly being updated and improved, so continue to check the site periodically.

Tricks for Making Bootstrap Sites Accessible

You can do a lot of things to make your Bootstrap CSS and components more accessible.

When you're building forms, you should use labels with the `<label>` tag for every input field. But if you don't want the labels to display, you can hide them with the `.sr-only` class as you learned in Hour 9.

With the `.has-feedback` class and the proper icon, Bootstrap fields will show a small icon on the right. But this is not accessible by itself. To make it accessible, you need to add both a text description of the icon and the `aria-describedby` attribute pointing to that description. Listing 22.2 shows that you can hide the description with the `.sr-only` or `.sr-only-focusable` classes.

LISTING 22.2   A Form Field with Icon Made Accessible

```
<div class="form-group has-success has-feedback">
  <label class="control-label" for="fullname">Fill in Your Full
  Name</label>
  <input type="text" class="form-control" id="fullname"
         aria-describedby="fullnameStatus" placeholder="Full Name">
  <span class="glyphicon glyphicon-ok form-control-feedback"
        aria-hidden="true"></span>
  <span id="fullnameStatus" class="sr-only">(success)</span>
</div>
```

When using the Bootstrap Glyphicons, you can make them more accessible by adding descriptive text, but you should also hide them with the `aria-hidden="true"` attribute. Modern AT devices will announce CSS-generated content, like the Glyphicons, unless you explicitly hide it. But make sure if the icon is meant to convey meaning that you provide alternative text.

The Bootstrap navigation menus and navbars can be made more accessible by putting the `role="navigation"` attribute on the parent container element. This tells AT that the contained list is navigation. Be sure you do not put it on the `<ul>` element itself because this would prevent the device from announcing the list. Also, it's best that the container element be an HTML5 `<nav>` element.

Give button groups a role of `group` and toolbars a role of `toolbar`. Then if you include an `aria-labeledby` attribute, AT devices will announce them correctly using the label.

You should always add the `role="dialog"` attribute to modals. Next, you should use the `aria-labelledby` attribute to point to the modal title. You also can describe your modal and use the `aria-describedby` attribute to point to that text. Use the `aria-hidden="true"` attribute to tell AT to skip the modal's DOM elements.

Collapsible elements use the `aria-expanded` attribute. If the collapsed element is closed by default, it should have the `aria-expanded="false"` attribute. If it is open by default using the `.in` class, it should have the `aria-expanded="true"` attribute set. You also should place the `aria-controls` attribute on an element that controls one collapsible element. It should point to the ID of the collapsible element.

Summary

In this hour you learned what accessibility is when it comes to websites. You learned about the WAI-ARIA and basic accessibility principles. You also learned about how Bootstrap is set up to be accessible as well as some techniques for making Bootstrap sites more accessible.

You learned why skip navigation is important and how to build it. You learned how to create nested headings so that assistive technology can create a table of contents for your pages, and you learned about color contrasts and some places where the contrast is not ideal in the Bootstrap theme.

Workshop

The workshop contains quiz questions to help you process what you've learned in this hour. Try to answer all the questions before you read the answers.

Q&A

Q. Are there laws about website accessibility?

A. This depends on the country and municipality where you live. But most countries have laws requiring government and public sector websites be accessible. Additionally, some private companies have been sued because their websites were inaccessible.

Ultimately, you should not be making your site accessible because of fear of a lawsuit. Instead, you should keep in mind that accessibility is just a way of including more customers—and the more customers you have, the more business your website will do.

Q. Is there any special way to make video accessible?

A. Just like with images and other multimedia content, you should always have a transcript of the speech in the video. If you have audible content, then you should include a description of those sounds. The best way to get a transcript is to type it yourself or pay someone to do it. However, you can use software tools that use speech recognition. Captions also make videos more accessible.

Quiz

1. Which is more important: accessible design or inclusive design?

    a. Accessible design.

    b. Inclusive design.

    c. They are both equally important.

    d. Neither, you should focus on usable design.

2. Which of the following is NOT something you should consider when working on accessibility?

    a. The design complexity

    b. The number of images

    c. Audio and video content

    d. Colors

3. What makes this HTML less accessible?

```
<a href="page.html">Click here</a> to read more...
```

    a. The link is too short to click on.

    b. The link URL is not descriptive enough.

    c. The link text is not descriptive enough.

    d. The link doesn't have a `role` attribute.

4. Which attribute points to the ID of a block of text that explains a nontextual element like a video?

    a. `aria-described`

    b. `aria-describedby`

    c. `aria-description`

    d. `aria-label`

5. Where do you place skip navigation?

    a. At the very bottom of the HTML document, just before the `</html>` tag.

    b. At the very bottom of the web page, just before the `</body>` tag.

    c. At the very top of the HTML document, just after the `<html>` tag.

    d. At the very top of the web page, just after the `<body>` tag.

6. Why do you need the `tabindex` attribute on in-page links?

    a. Because some browsers won't focus on in-page links without an explicit tab order.

    b. Because it is a required attribute for AT devices to read links.

    c. Because in-page links don't work without it.

    d. You don't need the `tabindex` attribute.

7. True or False: Using headlines in a numerical hierarchy enables AT to create tables of contents for web pages.

8. What are ways that color can cause accessibility issues?

    a. Colors that look similar to color blind people

    b. Colors with too low a contrast

    c. Colors that provide information with no fallback

    d. All of the above

9. Which Bootstrap class allows you to hide content from all but AT devices?

    a. `.at-only`

    b. `.at-visible`

    c. `.sr-only`

    d. `.sr-visible`

10. What does AT stand for?

    a. Accessible technology

    b. Assistive technology

    c. Assistive terminal

    d. Accessible terminal

Quiz Answers

1. c. They are both equally important.

2. b. The number of images is not important to an accessible design, as long as they have alternative text or a defined role for AT.

3. c. The link text "Click here" is not descriptive enough.

4. b. `aria-describedby`

5. d. At the very top of the web page, just after the `<body>` tag.

6. a. Because some browsers won't focus on in-page links without an explicit tab order.

7. True

8. d. All of the above.

9. c. `.sr-only`

10. b. Assistive technology

Exercises

1. Check how accessible your website is by testing the navigation using only your keyboard. Turn off or unplug your mouse, and then use the Tab key to browse the page and use the Enter key to select elements. Next, try to navigate through forms using the arrow keys. This is not a definitive accessibility test, but it can give you an idea of possible problems.

2. Find a color contrast analysis tool online, and test your site with it. A good one is the Colour Contrast Analyzer (http://www.paciellogroup.com/resources/contrastanalyser/), which has a Windows and a Macintosh version. It will test against recognized accessibility standards.

Using Less and Sass with Bootstrap

What You'll Learn in This Hour:

▶ What a CSS preprocessor is
▶ Some basic features of Less
▶ How to use Less to update your Bootstrap website
▶ Some basic features of Sass
▶ How to use Sass to update your Bootstrap website

Bootstrap ships with plain CSS, but it uses both Less and Sass CSS preprocessors. You can use them to adjust how your Bootstrap website looks. In this hour you learn the rudiments of these preprocessors and how to use them with Bootstrap.

What Is a CSS Preprocessor?

Many people often think that CSS doesn't need preprocessing, and for small sites that is probably the case. But preprocessors make doing large or extensive changes across websites much easier.

A *CSS preprocessor* is a language that extends CSS and adds traditional programming features such as variables, loops, if-statements, and so on. Then the code is compiled into standard CSS for use in web browsers. CSS preprocessors add a number of benefits you can't get with plain vanilla CSS:

▶ **Don't Repeat Yourself (DRY)**—CSS forces you to write many styles over and over and over. CSS preprocessors let you define variables and snippets of reusable code so you're not typing the same thing repetitively.

▶ **Cleaner CSS**—With a preprocessor, you can do things like nest styles and use variables to store information so the styles make more sense.

▶ **Faster updates**—If you have a large site with a large CSS file, it can take a long time to find every instance of a style. But if you have a color defined as a variable, you can change that variable in one place and know that every button, text block, or background using that color is changed.

▶ **Cross-browser support is built in**—Although browsers are getting better and better at using the official CSS properties, there are still a lot of prefixed styles and styles that require different syntax across browsers. Rather than looking it up every time and for every style that might need to be prefixed, preprocessors will add the correct CSS automatically.

▶ **Scripting**—Vanilla CSS is very straightforward and simple. If you need a style to be applied only in certain circumstances, you will need to apply it with JavaScript. But preprocessors include many scripting features to let you vary the styles right in the CSS.

▶ **Share CSS across sites**—You can store your CSS blocks as shareable library files to use in other pages or sites. You also can get snippets or libraries from other people on the Web so you don't have to build complex styles from scratch.

Using Less

Less is the preprocessor that Bootstrap comes with by default. By understanding some of the basic features of Less, you can improve how your Bootstrap sites work.

Less Features

Less has a lot of features that make it preferable to using plain CSS, including

▶ Variables

▶ Mixins

▶ Operations

▶ Functions

A variable in Less is a name/value pair defined at the beginning of a Less file. The variable or name starts with the @ sign and is separated from the value by a colon (:). You then use the variable anywhere you want that value to appear in your CSS. Listing 23.1 shows a simple color variable and how it's used in the Less CSS file.

LISTING 23.1 Simple Less Variable

```
@yellow: #F0E433;

p.pull-right {
  color: @yellow;
}
```

When you compile the Less, it will generate CSS replacing the variable @yellow with the color code #F0E433.

Mixins are a way of including properties from one rule set into another rule set. For example, say you had a class called .center-block that centered block elements. You could include that class in any other style property that you wanted centered as a mixin. Listing 23.2 shows a simple Less mixin.

LISTING 23.2   Simple Less Mixin

```
.centered-block {
  margin-right: auto;
  margin-left: auto;
}

h1 {
  width: 80%;
  .centered-block;
}
```

This will make all <h1> tags both 80% wide and centered.

Mixins also allow you to create default values but then change them in specific instances. For example, you might want to have most of your borders be 3px wide but want to be able to change the border width when needed. Listing 23.3 shows how to use a mixin with a default value and how to change that value.

LISTING 23.3   Parametric Mixins

```
.green-bordered(@borderwidth: 3px) {
  border: solid green @borderwidth;
}

p.pull-right {
  .green-bordered();
}

p.wide-border {
  .green-bordered(10px);
}
```

This sets the p.pull-right style to the default border width, while the p.wide-border style has a border width of 10px.

Operations allow you to adjust any number, color, or variable using math functions. This is straightforward when you're talking about adding or subtracting values from a number, like a width. But you also can add or subtract colors using color math.

For example, @width: 10px + 5; results in the output of 15px whenever the @width variable is used. Listing 23.4 shows how you might darken a color by a specific amount using operators and color math.

LISTING 23.4 Changing a Color with Operators

```
@yellow: #F0E433;
@other-yellow: @yellow - #666;

p.pull-right {
  color: @yellow;
  text-shadow: 2px 2px @other-yellow;
}
```

This creates a second color that is related to the first color mathematically. It is the base yellow color (#f0e433) minus #555 or #9b8f00.

Less has many built-in functions you can use to adjust your CSS. Functions change CSS from a simple text-based language to a programmable tool. There are functions to modify strings of text, list functions to get the length of a list or specific elements in a list, math functions for converting numbers, type functions to evaluate values, color functions for working with color definitions and channels, as well as functions to help you change and blend the colors you have.

In Bootstrap you see a lot of the color operation functions to create monochromatic color palettes. Listing 23.5 shows the grayscale color scheme created using just the color black (#000) and the lighten and darken functions.

LISTING 23.5 Create a Color Palette with Less Functions

```
@gray-base: #000;
@gray-darker: lighten(@gray-base, 13.5%);
@gray-dark: lighten(@gray-base, 20%);
@gray: lighten(@gray-base, 33.5%);
@gray-light: lighten(@gray-base, 46.7%);
@gray-lighter: lighten(@gray-base, 93.5%);

.black {
  color: @gray-base; // @gray-base
}
.darkest {
  color: @gray-darker; // @gray-darker
}
.dark {
  color: @gray-dark; // @gray-dark
}
.gray {
  color: @gray; // @gray
}
.light {
  color: @gray-light; // @gray-light
}
.lightest {
  color: @gray-lighter; // @gray-lighter
}
```

As you can see, these functions are much easier to understand what's going to happen than using the color math operations mentioned earlier. If you want to use a different color for @gray-base, you can—and this will generate a monochromatic color scheme for you.

There is much more to Less than can be covered in one hour. If you're interested in learning more, you should start with the Less home page at http://lesscss.org/.

Using Less with Bootstrap

It's easy to use Less with Bootstrap because Bootstrap is built with Less by default. You can use the source Less files, rather than the compiled CSS, to create your own custom designs. Or you can use the variables and mixins that are already created.

If you are going to work with the Less files, you will need to have a setup for compiling the CSS. Bootstrap recommends using Grunt (http://gruntjs.com/).

TRY IT YOURSELF ▼

Use Grunt to Install Bootstrap

To use Grunt to install Bootstrap, you must have Grunt on your system; you also need the Bootstrap package installed. This Try It Yourself walks you through the steps to install Grunt and use it to install a Bootstrap package on a Linux server using the command line. Some of these steps might require administrative (root) access to your server. If you do not have root access, contact your hosting provider to see whether it can install Grunt for you.

1. Verify that you have npm (Node Package Manager) installed on your server by typing

   ```
   which npm
   ```

   If you don't have npm installed, you must download and install it using root access. You can find npm at http://nodejs.org/download/.

2. As root, install the Grunt client:

   ```
   npm install -g grunt-cli
   ```

   If you don't have root access, you must contact your hosting provider to have it install the Grunt client.

3. Change to the directory where you want to install your Bootstrap files:

   ```
   cd bootstrap
   ```

4. Install Bootstrap using npm; you need to still be logged in as root or another admin user:

   ```
   npm install bootstrap
   ```

5. Change to the Bootstrap directory:

   `cd node_modules/bootstrap`

6. Run `npm install` in that directory. If you get an error message, edit the `package.json` file and remove the line that reads:

   `"grunt-sed": "twbs/grunt-sed#v0.2.0",`

   Then run `npm install` again.

You will then have five Grunt commands:

► `grunt dist`—Regenerates the CSS and JavaScript in the `dist/` directory. This will be the most common command you run.

► `grunt watch`—Watches the Less source and automatically recompiles the CSS when you make a change.

► `grunt test`—Runs tests on your files.

► `grunt docs`—Builds and tests the documentation assets.

► `grunt`—Builds and compiles everything and tests it all.

Be aware that this doesn't work on all Linux distributions in the same way. If you can't get Grunt installed, you should talk to your hosting provider for more help.

After you have Bootstrap and Grunt installed, you can edit the files in the source code directories—less/, js/, and fonts/—to reflect your website styles, scripts, and fonts. Use the grunt dist command to regenerate your distribution files.

CAUTION

Run Grunt Commands in the `less/` Directory

If you run the Grunt commands from within the distribution directory, all your files will be removed because there are no Less files to compile there. However, if you do this, don't panic. Just change to the less/ directory and rerun the Grunt command: `grunt dist`. This will re-create your entire distribution directory.

Using Sass

Sass (http://sass-lang.com/) is another CSS preprocessor that works with Ruby or inside applications such as Compass (http://compass.kkbox.com/). Many web designers are switching from Less to Sass because Sass offers more features than Less does. It also solves some problems that Less can have, such as when using the @ symbol for variable names.

But to use Sass with Bootstrap, you have to use a different port of Bootstrap that is adjusted to use Sass. You'll learn more about that in the section "Using Sass with Bootstrap."

Sass Features

Sass has many of the same features as Less, including

▶ Variables

▶ Mixins

▶ Operators

▶ Nesting

▶ Logic and loops

Variables work the same way in Sass as they do in Less. You replace long strings or commonly used code blocks with short words or phrases. The only difference is that Sass uses the $ rather than the @ sign to define variables. This has the advantage of not being confused with CSS @ declarations such as @media and @import. Plus, many other programming languages use the dollar sign as a variable identifier. Listing 23.1 showed a simple Less variable; Listing 23.6 shows you the same variable in Sass.

LISTING 23.6 Simple Sass Variable

```
$yellow: #F0E433;

p.pull-right {
  color: $yellow;
}
```

Mixins also work in a similar way to Less, but you define them with the @mixin directive. You can pass variables to your mixins as well. Then, you just include the mixin in the style(s) you need it in using the @include directive. Listing 23.7 shows how to use a simple Sass mixin.

LISTING 23.7 Simple Sass Mixin for Rounded Corners

```
@mixin border-radius($radius) {
  -webkit-border-radius: $radius;
  -moz-border-radius: $radius;
  -ms-border-radius: $radius;
  border-radius: $radius;
}

.box { @include border-radius(10px); }
```

Mixins are often used to deal with browser prefixes, and you can find lots of examples of useful Sass mixins online. A number of mixins are built right into Bootstrap. One of my favorites is the `@mixin size()` mixin. This mixin takes two parameters: width and height. You can then set your Bootstrap elements to specific sizes with just one line of code. Listing 23.8 shows how.

LISTING 23.8   Using the `size()` Mixin

```
.myImage {
  @include size (400px, 300px)
}
```

Like in Less, you can use operators to do math from within your Sass CSS files. You can perform many other operations, including these:

- **Number operations**—These are the standard math operations you're familiar with such as `width: 100px * 2px`.

- **Color operations**—You also can add, subtract, multiply, and divide colors just like in Less. These operations work on the red, green, and blue channels in turn.

- **String operations**—Use the + operator to concatenate strings together.

- **Boolean operations**—You can use the and, or, and not operators for Boolean values.

Nesting lets you include the rules for one property inside another. This makes the CSS appear more like it does in the HTML itself. Listing 23.9 shows a simple nested style and how it compiles into CSS.

LISTING 23.9   A Nested Style and the Resulting Compiled CSS

```
// Sass code
#main p {
  color: #000000;
  width: 98%;

  .highlight {
    background-color: #f1b161;
  }
}

// Output CSS
#main p {
  color: #000000;
  width: 98%;
}
main p .highlight {
  background-color: #f1b161;
}
```

You can extend styles so that they have the same styles as the previous style, but with some extra. For example, you might have a standard paragraph style with a font family and background color set. Say you want your `<aside>` tags to look the same as your paragraphs but with a border around them as well. Listing 23.10 shows how easily you can do this in Sass using the `@extend` directive.

LISTING 23.10   Using the `@extend` Directive

```
.standard {
  font-family: "Source Sans Pro";
  background-color: #e6b29a;
}
aside {
  @extend .standard();
  border: solid #efefef 1px;
}

// compiled CSS
.standard, aside {
  font-family: "Source Sans Pro";
  background-color: #e6b29a;
}
aside {
  border: solid #efefef 1px;
}
```

Although Less has some logic statements (`when`) and a little bit of looping, Sass has these features fully developed. It has `if/then/else` statements, `for` loops, and `each` loops. These work exactly how they work in other programming languages.

There is a lot more to Sass than can be covered in this book. If you are interested in learning more, I recommend starting with the Sass documentation online at http://sass-lang.com/documentation/file.SASS_REFERENCE.html.

Using Sass with Bootstrap

Bootstrap is built on Less, but there is an official Sass port on GitHub: https://github.com/twbs/bootstrap-sass. You can install Bootstrap with Sass support manually, but it's easiest to use a tool like CodeKit (https://incident57.com/codekit/), Compass.app (http://compass.kkbox.com/), or Prepros (https://prepros.io/) that will automatically watch your files and compile them when necessary. These tools make it easy to use Sass and Bootstrap without having to worry about command-line tools or installing Ruby. You just download the program of your choice, install it, and set it to check the directories where you are storing your SCSS files.

▼ TRY IT YOURSELF

Setting Up a Bootstrap Sass Project with Compass.app

When you use Compass.app to build and maintain a Bootstrap Sass project, all you do is set up your project once in Compass.app. Then every time you edit the Sass or SCSS files, Compass.app will automatically compile the CSS for you. When you're ready to launch, you simply upload the CSS and HTML files to the web server as you normally would. Plus, by using Compass, you won't have to maintain your CSS vendor prefixes or make your sprite images by hand:

1. Download and install Compass.app from KKBox.com (http://compass.kkbox.com/). Note that this is not free software, but it is inexpensive ($10) and there are ways to get it for free.

2. Create a Compass Bootstrap Sass project by going to your Compass.app menu item and selecting Create Compass Project. Scroll to the bootstrap-sass-* and select project.

3. Save the project files in the directory of your web project. If you don't have a directory for the new project, you can create one now. Compass.app will then create all the files and folders for your project.

4. Edit the SCSS files in the `sass/` directory with your favorite text or web editor. You can change the default Bootstrap variables by editing the `_bootstrap-variables.scss` file. Be sure to remove the initial `//` to uncomment any variables you want to edit.

5. You can force Compass.app to recompile your CSS at any time by clicking the Clean and Recompile option in the menu.

After you have Sass installed and a compiler set up, you can edit your SCSS or Sass files exactly as you want to make your site as unique as you like.

Summary

This hour you learned about how to customize your Bootstrap installations using the Less and Sass preprocessors. You learned some of the basic features of both Less and Sass and learned how to get started editing Less and Sass files. You also learned how to compile them to make your Bootstrap sites look amazing.

Workshop

The workshop contains quiz questions to help you process what you've learned in this hour. Try to answer all the questions before you read the answers.

Q&A

Q. Where can I go to learn more about Less and Sass?

A. The first place you should start with both languages is their home pages: http://lesscss. org/ for Less and http://sass-lang.com/ for Sass. If you prefer learning from video courses, Lynda.com (http://www.lynda.com/) has several courses on both Less and Sass.

Q. If I customize the Bootstrap variables, what will happen when a new version of Bootstrap comes out?

A. It depends on how you update your files. If you edited the bootstrap CSS files directly (or the corresponding Less or Sass files), then those files might be overwritten when you load a new version. Your best solution is to create all your changes in a separate file and then load the file using include files with Less or Sass. Listings 23.11 and 23.12 show how you can include files using Less and Sass, respectively.

LISTING 23.11 Including Files with Less

```
// import the file my-bootstrap-styles.less
@import "my-bootstrap-styles.less"
```

If the file to import is in a different directory, include that inside the quotes.

LISTING 23.12 Including Files with Sass

```
// import the file _my-bootstrap-styles.scss
@import "my-bootstrap-styles"
```

The main difference between the two files is that the Sass file does not need to include the file extension (scss) or the underscore at the beginning of the filename.

Quiz

1. Why would you want to use a CSS preprocessor?

    a. To stay current with design trends

    b. To improve the flexibility of your website

    c. To manage large or extensive changes on a website

    d. To keep your CSS looking nice

2. What is a good reason to not use a CSS preprocessor?

    a. When you have a small site with little or no CSS.

    b. When you don't want to compile CSS.

    c. When you don't know the languages.

    d. You should always use a CSS preprocessor.

3. What does DRY mean in the context of CSS preprocessors?

    a. Don't Reuse sYmbols

    b. Don't Repeat Yourself

    c. Do Repeat Yourself

    d. Nothing

4. True or False: CSS preprocessors let you use `if/then/else` statements.

5. How do you define a Less variable?

    a. The at sign (@)

    b. The caret symbol (^)

    c. The dollar sign ($)

    d. The hash symbol (#)

6. True or False: Less mixins do not accept parameters.

7. Which of the following is a Less function?

    a. `light`

    b. `lighter`

    c. `lighten`

    d. `lightest`

8. How do you define a Sass variable?

    a. The at sign (@)

    b. The caret symbol (^)

    c. The dollar sign ($)

    d. The hash symbol (#)

9. True or False: The following code is valid in Sass:

```
font-family: sans- + "serif";
```

10. Which preprocessor allows for `for` loops: Less or Sass?

    a. Both

    b. Less

    c. Sass

    d. Neither

Quiz Answers

1. c. To manage large or extensive changes on a website

2. a. When you have a small site with little or no CSS

3. b. Don't Repeat Yourself

4. True

5. a. The at sign (@)

6. False

7. c. `lighten`

8. c. The dollar sign ($)

9. True

10. c. Sass

Exercises

1. Install Grunt on your machine and set it up to watch a Bootstrap Less directory. Then make some changes to the Less files, and compile them using the Grunt command `grunt dist`. After you have compiled your CSS, upload it to your web server to view it live.

2. Install Compass on your machine and use it to install Sass. Then create a new Bootstrap with Sass folder, and build a new website using what you've learned in previous hours.

Going Further with Bootstrap

Bootstrap is a powerful tool for web designers to create responsive websites and nice-looking applications. But there is so much more you can do with Bootstrap when you know how to use it properly.

In this hour, you learn some of the tools, add-ons, and websites you can use to improve your Bootstrap websites. You also learn about editors that have Bootstrap built in as well as how to use Bootstrap in tools such as WordPress. Finally, you find out where you can go to extend your knowledge of Bootstrap.

Bootstrap Editors

The best editors for Bootstrap are the ones you already use and are comfortable with. But it can be nice to have some of the components and features built right into your editor. Plus, some of the Bootstrap editors out there offer the ability to easily and quickly customize the Bootstrap theme, so your websites will look amazing from the outset.

Web Editors

A number of online and offline web editors exist that you can use to create websites with Bootstrap. When you're looking for an editor for your Bootstrap websites, here are a few things you should look for:

▶ Is it a visual editor or a code editor (or both)?

▶ Can you modify the CSS, add components, and include Bootstrap JavaScript elements?

▶ Can you edit the HTML, CSS, and JavaScript beyond the Bootstrap features?

▶ Can you edit and compile Less or Sass files?

▶ Can you use existing themes or style sheets?

It is tempting to evaluate editors based solely on price. Because Bootstrap is a free framework, spending money on an editor might seem foolish. But some of the best editors that have the most features also come with a cost. You can start with the free ones, but you might decide down the road that spending a little money can make a difference in improving your websites.

One of the biggest differences in web editors that support Bootstrap is whether they are online (a web-based application) or offline (a software-based application). Software applications work even if your computer is offline but require that you have a specific operating system to run them. Web applications run in the cloud and so require an Internet connection, but they can be run on any device that supports the Web.

I have found six online applications you can use to create and edit Bootstrap pages:

▶ **Bootply (http://bootply.com/)**—Offers a free service as well as paid options for hosting your applications. They store the creations as snippets (they call them "plys") that are shared with the entire Bootply community. This is a popular option.

▶ **BootTheme (http://www.boottheme.com/)**—It's free for noncommercial use for up to five applications and starts at $9.99 per month for commercial applications. Don't be fooled by the name, though. BootTheme is both a theme editor and a visual web page editor. It does not include Bootstrap JavaScript.

▶ **Brix.io (http://brix.io/)**—Starts at $4.90 per month with a 14-day free trial. It includes a visual editor and some nice premade templates, and it exports your sites as HTML5, but with some custom classes.

▶ **Divshot (https://divshot.com/)**—Offers a free basic account, with more advanced accounts starting at $20 per month. This is an application development platform that includes a command-line interface along with Bootswatch themes, a visual editor, and a code editor. You also can edit and compile Less from within Divshot.

▶ **Jetstrap (https://jetstrap.com/)**—Starts at $16 per month and does not have a free trial. It is a visual editor and supports Bootstrap base CSS and components; it does not support the Bootstrap JavaScript features.

▶ **LayoutIt (http://www.layoutit.com/)**—A free online layout builder. It creates the base layout for your Bootstrap website for you to download as HTML and CSS; then you can open those files in another editor to do the final edits.

I also found four offline applications that support creating and editing Bootstrap websites:

▶ **BootUI (http://www.bootui.com/)**—BootUI is available for Windows and Macintosh for $49.95. It is a Bootstrap template editor that comes with several nice-looking templates. The creators will add more templates from your mock-ups if you ask for them.

▶ **Dreamweaver CC 2015 (http://www.adobe.com/products/dreamweaver.html)**—Dreamweaver is available for Windows and Macintosh starting at $19.99 per month. The 2015 version of Dreamweaver comes with a Bootstrap integration built in. If you open a site that is already built with Bootstrap, Dreamweaver CC 2015 detects that and opens the Bootstrap Components insert panel. You also can create new Bootstrap pages using pre-built templates or a blank page.

▶ **Pinegrow (http://pinegrow.com/)**—Pinegrow is available for Windows, Macintosh, and Linux starting at $49 with a 30-day money-back guarantee. It includes all the Bootstrap CSS, components, and JavaScript in a visual editor. You also can edit the code directly.

▶ **Pingendo (http://www.pingendo.com/)**—Pingendo is available for Windows, Macintosh, and Linux for free. It is a visual editor that allows you to edit the CSS and components, but not the JavaScript.

Theme Builders and Customizers

You also can find tools that let you edit Bootstrap themes and customize your Bootstrap websites. These tools typically help you adjust the colors and fonts in Bootstrap, but some also help you create buttons or adjust the grid to really make your Bootstrap site different.

Theme builders are really helpful for people who already have a web editor that they are happy with, but who find editing the Bootstrap CSS or Less and Sass files difficult. With these tools, you can create your Bootstrap site and then drop in the CSS generated by the theme builder to modify the look and feel to suit yourself.

I found six online free theme builders you can use to create Bootstrap themes in different ways:

▶ **Bootstrap Magic (http://pikock.github.io/bootstrap-magic/app/#!/editor)**—Bootstrap Magic lets you customize every aspect of Bootstrap CSS and components using a preview engine to show how your changes will look as you apply them. It uses Less variables in the form itself and autocompletes to help you get the correct structure. It also has color pickers to help you get nice colors. Plus, you can preview typographic changes. It exports as both Less and CSS (both minified and standard). You can import existing themes as Less variables.

▶ **Bootswatchr (http://bootswatchr.com/)**—Bootswatchr uses the Less variables to help you see how your theme will look as you build it. It is especially useful if you are already familiar with the Less variables file and you just want to see how your changes will look in a live environment. It previews both color and typography. It exports as CSS and minified CSS, and you can import Bootswatches you've previously saved.

▶ **Lavish (http://www.lavishbootstrap.com/)**—Lavish takes an online photo and creates a Bootstrap color scheme from it. You can then adjust the colors to match exactly how you want your page to look. You can export your colors as either CSS or Sass.

▶ **Paintstrap (http://paintstrap.com/)**—Paintstrap uses ColourLovers and Kuler codes to generate color schemes for Bootstrap pages. It doesn't seem to support the new Adobe Color CC, which is what Kuler has changed into, but it still works great with ColourLovers. You can export your color theme as CSS, both minified and standard, as well as Less variables.

▶ **Style Bootstrap (http://stylebootstrap.info/)**—Style Bootstrap enables you to change settings and watch the preview change as you type. It also offers a "random" button to generate some truly hideous-looking color schemes. You can export your color schemes as CSS.

▶ **Twitter's Bootstrap 3 Navbar Generator (http://twitterbootstrap3navbars.w3masters. nl/)**—The name makes it appear to be a generator, but it is really a theme generator specifically for navbars. You can generate specific colors and styles for your navbars and then export the CSS for use in your Bootstrap pages.

I also found two generators that will help you generate the Bootstrap HTML code for forms and buttons:

▶ **Form Builder for Bootstrap (http://bootsnipp.com/forms)**—This is a drag-and-drop form builder that can help you quickly create a nice-looking Bootstrap form. It doesn't style the form for you (use the previously listed theme builders for that). Instead, it shows you how your form would look with drag-and-drop form elements. The only thing it doesn't do is generate HTML5 form fields such as `email` and `url`.

▶ **Twitter Bootstrap Button Generator (http://www.plugolabs.com/twitter-bootstrap-button-generator-3/)**—This generator helps you generate Bootstrap buttons. You can choose the button text and style, and even decide whether it has an icon on it and, if so, which icon to use.

Using Bootstrap in WordPress

Bootstrap is a great framework for building websites, but what if you want to use a content management system like WordPress? WordPress offers a lot of useful features for websites, including blogs and content management. But it doesn't use Bootstrap out of the box.

To use Bootstrap with WordPress, you have three options:

- ▶ Use a plugin to extend Bootstrap to WordPress.

- ▶ Find a WordPress theme that uses Bootstrap.

- ▶ Build your own theme using Bootstrap.

Using a WordPress Plugin

If you search the WordPress plugins site (https://wordpress.org/plugins/), you will find a lot of plugins that offer to extend your WordPress site with Bootstrap. The two that I particularly like are Easy Bootstrap Shortcode (https://wordpress.org/plugins/easy-bootstrap-shortcodes/) and WordPress Twitter Bootstrap CSS (https://wordpress.org/plugins/wordpress-bootstrap-css/).

Both offer shortcodes for using Bootstrap, but I find the Easy Bootstrap Shortcode to be a bit easier to use. It doesn't include all the same components as WordPress Twitter Bootstrap CSS, but it has all the shortcodes included in the tinyMCE editor, as shown in Figure 24.1.

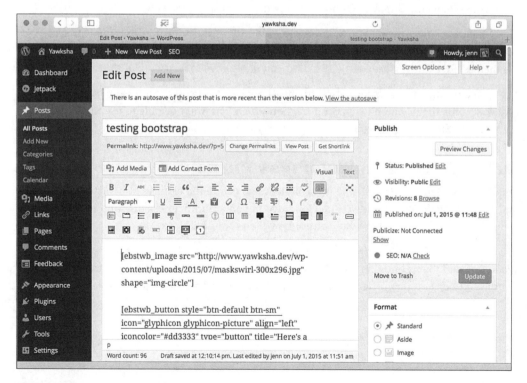

FIGURE 24.1
Shortcodes in the visual editor with Easy Bootstrap Shortcode plugin.

The benefit to using a plugin to add Bootstrap to your WordPress sites is that you don't need to make any major changes to your site. You can get some of the benefits of Bootstrap (like using accordions or decorating your buttons with icons) without a lot of effort.

The drawback is that plugins provide a limited selection of Bootstrap. You're not getting the full Bootstrap experience because the pages themselves are not built in the Bootstrap framework. Instead, you have sections of your pages that use Bootstrap and sections that do not.

Finding a Bootstrap Theme for WordPress

Just like with plugins, if you search the WordPress Theme Directory (https://wordpress.org/themes/search/bootstrap/), you'll find a lot of themes using Bootstrap. There are basic themes that give you the full Bootstrap experience, with little customization, and there are some much fancier themes with a lot more options.

As with all themes, you should test the theme you're planning to use extensively before launching with it—you might discover some major problems. For instance, one theme I considered looked gorgeous on large screens but was completely illegible on an iPhone 6. This was surprising because Bootstrap handles multiple breakpoint responsive designs well.

The benefit to using a theme to add Bootstrap to WordPress is that you get all the extras of Bootstrap, including the grid system and all the CSS, components, and JavaScript. Plus, many of the themes have already been customized, so you don't have to worry about that.

The problem with using someone else's theme, though, is that you are limited to what they give you. If the theme doesn't support accordions, then you won't be able to use them. Or if the theme isn't as responsive as you would like, then you have to live with that, too.

Building Your Own WordPress Theme

Building your own Bootstrap theme is the best way to get all the features of Bootstrap you need and want into your WordPress website. You have the opportunity to use any and every Bootstrap feature in your website because you control how the website will look.

The challenge is that building your own theme requires that you know a lot more about WordPress and theme building than I can go into in this hour. A good place to start is the WordPress Codex on theme development (http://codex.wordpress.org/Theme_Development).

Build a Basic WordPress Theme with Bootstrap

Learn how to create a basic Bootstrap blog theme using the Bootstrap blog template page (http://getbootstrap.com/examples/blog/). To do this, you will need access to your WordPress site file system (either by command line or through an admin tool like cPanel):

1. Create a directory in your `wp-content/themes` directory called `my-bootstrap-theme/`.

2. Open a new file called `style.css`, and save it to the same directory.

3. Add the WordPress stylesheet header in Listing 24.1 to the `style.css` file. Change the values to reflect your theme.

LISTING 24.1   The `style.css` Document for a Bootstrap WordPress Theme

```
/*
Theme Name:    My Bootstrap Blog Theme
Theme URI:     http://example.com/my-bootstrap-theme
Author:        my name
Author URI:    my URL
Description:   This is a theme based off the Bootstrap Blog
               template:
               http://getbootstrap.com/examples/blog/.
Version:       1.0
License:       Attribution-ShareAlike 3.0 Unported
License URI:   http://creativecommons.org/licenses/by-sa/3.0/
Tags:          blue, light, two columns, responsive, bootstrap
*/
```

4. Grab the HTML from the Bootstrap Blog template, and save it as `index.php` in the my-bootstrap-theme folder.

5. Create a simple image 300 × 225 to represent your theme. Save it as `screenshot.png` in the same directory.

6. Download Bootstrap and unzip the files in the theme directory. You should have a directory structure like this:

```
my-bootstrap-theme/
├── bootstrap/
      ├── css/
      ├── js/
      └── fonts/
```

7. Go to your WordPress admin area and navigate to the Appearance > Themes area. You will see your new theme ready to set up as in Figure 24.2.

FIGURE 24.2
My Bootstrap theme available in WordPress.

8. Activate your theme, but be aware that you won't be able to see anything in it because there are no WordPress tags in the theme.

NOTE

Live Sites Let Your Theme Be Seen

If you are working on a live site, you should install a plugin such as Theme Test Drive (https://wordpress.org/plugins/theme-test-drive/) to ensure that only you can see the new theme until you are ready to take it live. This plugin lets the administrator of the blog see new themes until they are ready to go live, while regular visitors see the old theme.

9. Open the `index.php` file in your favorite editor, and cut everything in the file from the first line to line 53 that reads `<div class="row">`. Paste that into a new file called `header.php`. Listing 24.2 shows you what `header.php` should look like. Replace that text with `<?php get_header(); ?>`.

LISTING 24.2   Your Initial `header.php` File

```
<!DOCTYPE html>
<html lang="en">
  <head>
    <meta charset="utf-8">
```

```
    <meta http-equiv="X-UA-Compatible" content="IE=edge">
    <meta name="viewport" content="width=device-width,
    initial-scale=1">
    <!-- The above 3 meta tags *must* come first in the head; any
    other head content must come *after* these tags -->
    <meta name="description" content="">
    <meta name="author" content="">
    <link rel="icon" href="../../favicon.ico">

    <title>Blog Template for Bootstrap</title>

    <!-- Bootstrap core CSS -->
    <link href="../../dist/css/bootstrap.min.css" rel="stylesheet">

    <!-- Custom styles for this template -->
    <link href="blog.css" rel="stylesheet">

    <!-- Just for debugging purposes. Don't actually copy these 2
    lines! -->
    <!--[if lt IE 9]><script
src="../../assets/js/ie8-responsive-file-warning.js"></script>
    <![endif]-->
    <script src="../../assets/js/ie-emulation-modes-warning.js">
    </script>

    <!-- HTML5 shim and Respond.js for IE8 support of HTML5
    elements and media queries -->
    <!--[if lt IE 9]>
<script
  src="https://oss.maxcdn.com/html5shiv/3.7.2/html5shiv.min.js">
</script>
<script src="https://oss.maxcdn.com/respond/1.4.2/respond.min.js">
</script>
    <![endif]-->
  </head>

  <body>

    <div class="blog-masthead">
      <div class="container">
        <nav class="blog-nav">
          <a class="blog-nav-item active" href="#">Home</a>
          <a class="blog-nav-item" href="#">New features</a>
          <a class="blog-nav-item" href="#">Press</a>
          <a class="blog-nav-item" href="#">New hires</a>
          <a class="blog-nav-item" href="#">About</a>
```

```
      </nav>
    </div>
  </div>

  <div class="container">

    <div class="blog-header">
      <h1 class="blog-title">The Bootstrap Blog</h1>
      <p class="lead blog-description">The official example
      template of creating a blog with Bootstrap.</p>
    </div>

    <div class="row">
```

10. In the `index.php` file, go to the bottom and cut everything from the `<footer>` to the bottom of the document. Paste that into a file called `footer.php`, and replace that text with `<?php get_footer(); ?>`. Listing 24.3 shows what will be in the `footer.php` file.

LISTING 24.3   The `footer.php` File

```
<footer class="blog-footer">
  <p>Blog template built for <a href="http://getbootstrap.com">
  Bootstrap</a> by <a href="https://twitter.com/mdo">@mdo</a>.
  </p>
  <p>
    <a href="#">Back to top</a>
  </p>
</footer>



  <!-- Bootstrap core JavaScript
  ================================================== -->
  <!-- Placed at the end of the document so the pages load
  faster -->
<script
src="https://ajax.googleapis.com/ajax/libs/jquery/1.11.3/
jquery.min.js"></script>
<script src="../../dist/js/bootstrap.min.js"></script>
  <!-- IE10 viewport hack for Surface/desktop Windows 8 bug -->
  <script src="../../assets/js/ie10-viewport-bug-workaround.js">
  </script>
  </body>
</html>
```

11. Inside the `header.php` file, replace the two CSS links with one line of HTML: `<link href="<?php bloginfo('stylesheet_url');?>" rel="stylesheet">`. This points to the `style.css` file that WordPress uses. While you're in there, remove the two lines about debugging that point to files in `../../assets/js/` because you won't need them.

12. Edit the `style.css` file to point to the Bootstrap CSS by adding an `@import` line at the top, just below the theme information comment:

```
@import url('bootstrap/css/bootstrap.css');
```

13. Edit the `header.php` file to add jQuery and the WordPress header information. Right above the closing `</head>` tag, add the following two lines:

```
<?php wp_enqueue_script("jquery"); ?>
<?php wp_head(); ?>
```

These ensure that jQuery is loaded in the `<head>` and there is a hook for WordPress plugins to add to the `<head>` as well.

14. Edit your `footer.php` file and remove everything between the `</footer>` and the `</body>` tags. Replace that text with: `<?php wp_footer(); ?>`. This will add a hook for WordPress to add any scripts or plugins to the footer that you might add to your installation later.

15. Create a `functions.php` file to add the Bootstrap JavaScript to your pages. It will look like Listing 24.4.

LISTING 24.4   The `functions.php` File

```php
<?php

function bootstrap_jquery_scripts() {
    // Register the script
    wp_register_script( 'my-script', get_template_directory_uri() .
    '/bootstrap/js/bootstrap.js', array( 'jquery' ) );
    // Enqueue the script:
    wp_enqueue_script( 'my-script' );
}
add_action( 'wp_enqueue_scripts', 'bootstrap_jquery_scripts' );

?>
```

16. Edit your `index.php` file so that it looks like Listing 24.5. This will get the WordPress Loop into the file, and you can start really playing with the theme.

LISTING 24.5   The Final `index.php` File

```php
<?php get_header(); ?>

        <div class="col-sm-8 blog-main">
<?php if ( have_posts() ) : while ( have_posts() ) : the_post(); ?>
```

```
        <div class="blog-post">
          <h2 class="blog-post-title"><?php the_title(); ?></h2>
          <p class="blog-post-meta"><?php the_time('F jS, Y'); ?>
            by <?php the_author_posts_link(); ?></p>

          <?php the_content(); ?>

        </div><!-- /.blog-post -->
<?php endwhile; else: ?>
    <p><?php _e('Sorry, no posts matched your criteria.'); ?></p>
<?php endif; ?>

        <nav>
          <ul class="pager">
            <li><a href="#">Previous</a></li>
            <li><a href="#">Next</a></li>
          </ul>
        </nav>

      </div><!-- /.blog-main -->

      <div class="col-sm-3 col-sm-offset-1 blog-sidebar">
        <div class="sidebar-module sidebar-module-inset">
          <h4>About</h4>
          <p>Etiam porta <em>sem malesuada magna</em> mollis
          euismod. Cras mattis consectetur purus sit amet
          fermentum. Aenean lacinia bibendum nulla sed
          consectetur.</p>
        </div>
        <div class="sidebar-module">
          <h4>Archives</h4>
          <ol class="list-unstyled">
            <li><a href="#">March 2014</a></li>
            <li><a href="#">February 2014</a></li>
            <li><a href="#">January 2014</a></li>
            <li><a href="#">December 2013</a></li>
            <li><a href="#">November 2013</a></li>
            <li><a href="#">October 2013</a></li>
            <li><a href="#">September 2013</a></li>
            <li><a href="#">August 2013</a></li>
            <li><a href="#">July 2013</a></li>
            <li><a href="#">June 2013</a></li>
            <li><a href="#">May 2013</a></li>
            <li><a href="#">April 2013</a></li>
          </ol>
        </div>
```

```
        <div class="sidebar-module">
          <h4>Elsewhere</h4>
          <ol class="list-unstyled">
            <li><a href="#">GitHub</a></li>
            <li><a href="#">Twitter</a></li>
            <li><a href="#">Facebook</a></li>
          </ol>
        </div>
      </div><!-- /.blog-sidebar -->

    </div><!-- /.row -->

  </div><!-- /.container -->

<?php get_footer(); ?>
```

17. Next, you need to add a little CSS to your `style.css` file to get the footer to look correct:

```
.blog-footer {
  padding: 40px 0;
  color: #999;
  text-align: center;
  background-color: #f9f9f9;
  border-top: 1px solid #e5e5e5;
}
```

18. Finally, upload your theme and test it in a web browser.

Be aware that this theme is not complete. You will have to do more work to add the dynamic sidebar and the navigation links and customize it more. You can read more about how to do that on the WordPress website. If you want to download a copy of all the theme files as they are listed here, you can get them from my website: http://www.html5in24hours.com/?attachment_id=1177.

Extending Bootstrap with Third-Party Add-ons

One of the great things about Bootstrap is how many people use it. When a lot of people use something, they often create things to improve it. This is definitely the case with Bootstrap. Bootstrap developers have created many themes and plugins you can use to improve and enhance your Bootstrap website.

Bootstrap Themes

Themes are the first place to look when deciding to upgrade your Bootstrap website. When you're looking for themes, the cheapest ones are going to just adjust the Bootstrap colors and sometimes typography. This type of theme also has the advantage of simply dropping in the new CSS or Less files and working with your pages. Bootswatch (http://bootswatch.com/) offers 16 free Bootstrap themes that you simply download and plug into your website.

However, to really make your Bootstrap site look different from other Bootstrap websites, you'll need to get a new template as well as a theme. A template can help you see the potential for your websites while not having to come up with the design all by yourself. Bootstrap Zero (http://www.bootstrapzero.com/) offers hundreds of free templates. It claims to be the largest open source free Bootstrap template collection, and I tend to agree. The site offers a lot of great color themes and templates to change the entire look of your website.

The place you're going to get the most bang for your buck, though, is with the professional templates. Before you stop reading, you should realize that "professional" does not have to mean "expensive." The site {wrap}bootstrap (https://wrapbootstrap.com/) offers hundreds of low-cost Bootstrap templates and themes. In fact, the most expensive themes I saw there cost less than $50. That $50 gets you dozens of different layouts, lots of color and typography options, and even WordPress support. Plus, most of the themes I saw cost less, often around $20, with some as low as $4. For just a few dollars, you can get all the benefits of Bootstrap with zero design work.

Bootstrap Plugins

Plugins are another great way to extend your Bootstrap website. These can give your site extra features or dress up the already nice-looking Bootstrap features.

Some things you can add with plugins include

- ▶ **Lightboxes**—Make your image galleries stand out more with Lightbox effects to open up thumbnails and provide more information. The Bootstrap Lightbox (http://www.jasonbutz.info/bootstrap-lightbox/) plugin makes it easy.

- ▶ **Photo Galleries**—More than just one image, a photo gallery plugin can help you manage the photos on your site pages and give them a great-looking display. The Bootstrap Image Gallery (http://blueimp.github.io/Bootstrap-Image-Gallery/) plugin even includes touch navigation.

- ▶ **Notifications and dialog boxes**—There are lots of plugins to make the Bootstrap modals (refer to Hour 15, "Modal Windows") even easier to build and maintain. Bootbox.js (http://bootboxjs.com/) is a popular plugin, but I really like Bootstrap Growl (https://github.com/ifightcrime/bootstrap-growl) because of all the customizations.

▶ **Web form helpers**—Web forms can be difficult to make, and even though Bootstrap makes it easier, a lot of form features are still left out. One of my favorites is the Bootstrap Form Helper (http://bootstrapformhelpers.com/). This collection of plugins offers several helpers to add things like date pickers, number inputs, and select elements into your forms. This plugin is not free for commercial use, and the cost starts at $12 per year. Other form plugins are available, but these are easy to use and quick to install.

There are lots more Bootstrap plugins than I can list. But one resource you can start with is the Big Badass List of 319 Useful Twitter Bootstrap Resources (http://www.bootstraphero.com/the-big-badass-list-of-twitter-bootstrap-resources). This list has more than just plugins, and it's a great place to start extending your Bootstrap site.

The Bootstrap Community

There is so much you can do with Bootstrap, it can be very overwhelming. Keep in mind, though, that thousands of other people just like you are getting started using this great framework. There are lots of online resources and places you can go to get help, but my three favorites are

▶ **Twitter Bootstrap Community Forums (http://www.twitterbootstrap.net/forum/forum.php)**—This is an online forum where you can ask and answer questions about Bootstrap. I wish it had a Sass section as well, but many of the questions I've had about Bootstrap have been answered there before I had to post a question myself.

▶ **Stack Exchange (http://stackoverflow.com/questions/tagged/twitter-bootstrap)**—This is just a list of questions about Bootstrap, but you will be surprised how many of your questions have already been answered here.

▶ **Reddit (http://www.reddit.com/r/bootstrap)**—The Reddit community is there for you if you have a problem with your Bootstrap implementation. Just be sure to include your page URL so that the feedback you get is useful, rather than just snarky.

Reaching out to the many other people who use Bootstrap on a daily basis is a great way to learn more about this framework. And remember, if you want to ask me a question, you can always contact me on my website http://htmljenn.com/. I am part of the Bootstrap community, too.

Beautiful Bootstrap Websites

You can create beautiful websites using the Bootstrap framework. You can find lots of examples of them on the website Built with Bootstrap (http://builtwithbootstrap.com/). You can even submit your own sites there now that you've finished this book. Figures 24.3–24.6 show a few sites you might be surprised were built with Bootstrap.

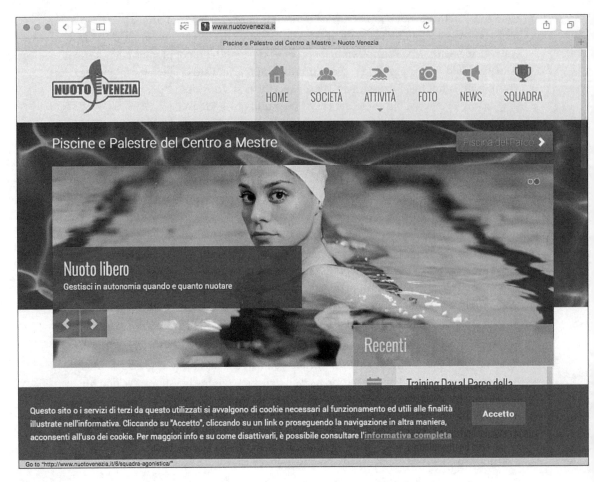

FIGURE 24.3
Nuoto Venezia (http://www.nuotovenezia.it).

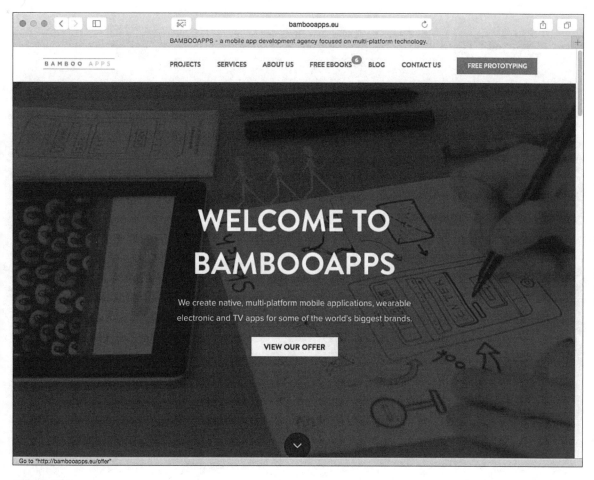

FIGURE 24.4
Bamboo Apps (http://bambooapps.eu/).

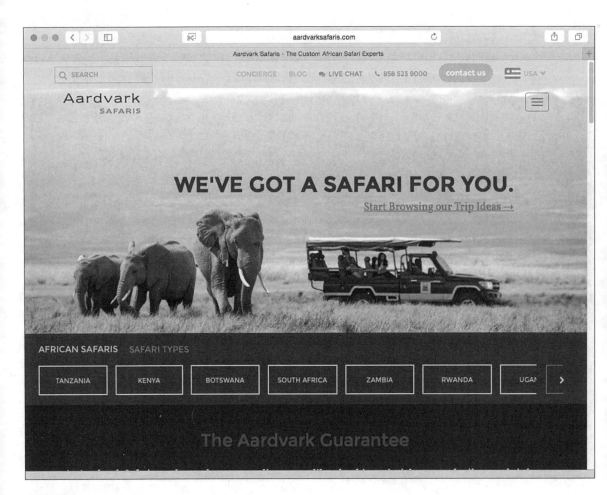

FIGURE 24.5
Aardvark Safaris (http://www.aardvarksafaris.com).

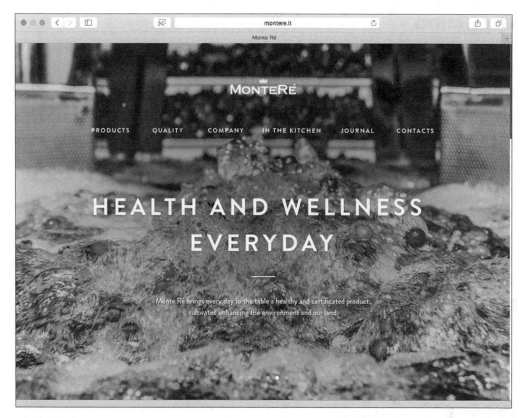

FIGURE 24.6
MonteRé (http://www.montere.it/?lang=en).

Bootstrap is a great framework, and you can build great sites with it.

Summary

This hour showed you some of the many ways you can extend your Bootstrap knowledge to create amazing Bootstrap websites. You learned about editors and tools to help you build and maintain your websites. You also learned about how to add Bootstrap to a content management system: WordPress.

This hour gave you some ideas for where to find third-party add-ons to improve your sites. You learned where to find free and professional themes, templates, and plugins and where to find other Bootstrap developers to get help and inspiration, as well as some beautiful Bootstrap websites to help you design your next amazing creation.

Workshop

The workshop contains quiz questions to help you process what you've learned in this hour. Try to answer all the questions before you read the answers.

Q&A

Q. What is the difference between a theme and a template?

A. In the Bootstrap world, a *theme* refers to the Less variables you can modify to make your website use different colors and typography. A *template* includes all the files you need to build a custom website. This usually means the HTML for one or more pages of a website, custom CSS, and sometimes JavaScript files.

Q. Can I use jQuery plugins if I can't find a Bootstrap plugin I like?

A. The world of Bootstrap plugins is huge and contains so many things you can add to your sites. But, if you can't find the perfect Bootstrap plugin to use, you can always use jQuery plugins because Bootstrap requires jQuery to run.

Quiz

1. Which of these is not something you need to look for in a Bootstrap web editor?

    a. Does it edit the Bootstrap CSS?

    b. Does it allow for visual editing?

    c. Can it compile Less or Sass?

    d. Will it add Bootstrap components?

2. When should you use a theme builder?

    a. When you don't know what your Bootstrap site should look like

    b. When you have a color palette but don't have the HTML for a layout

    c. When you have a color palette and the HTML for a layout

    d. When you have a complete template

3. Why should you use a WordPress plugin to incorporate Bootstrap?

    a. You should not use a WordPress plugin.

    b. Because they offer the strongest integration into WordPress.

    c. So that you get a better theme.

    d. So you don't have to make any big changes to your site theme.

4. True or False: Testing a Bootstrap theme is not important.

5. Which of the following is not true about building a Bootstrap WordPress theme?

    a. You can customize it to suit your site's needs.

    b. It is easy and quick to do.

    c. It lets you incorporate any of the Bootstrap components or plugins you want.

    d. You need to know a lot about WordPress to do it.

6. True or False: There are both online and offline Bootstrap editors.

7. Which of these is the best way to quickly change the look of a Bootstrap site?

    a. Design and build the HTML and CSS yourself.

    b. Add custom plugins.

    c. Hire a designer to build the HTML and CSS.

    d. Use a prebuilt theme or template.

8. What is the difference between a plugin and an add-on?

    a. There is no difference between a plugin and an add-on. They are two ways of saying the same thing.

    b. A plugin is completely finished, while an add-on requires work by the developer.

    c. A plugin works only with jQuery, while an add-on will work with any JavaScript.

    d. A plugin is specific to Bootstrap, while an add-on can work in any framework.

9. True or False: There are a lot of people using Bootstrap.

10. True or False: All Bootstrap sites look the same.

Quiz Answers

1. a. Does it edit the Bootstrap CSS? You don't want to edit the Bootstrap CSS. You should edit an alternate style sheet to add your changes.

2. c. When you have a color palette and the HTML for a layout. Theme builders let you define your Bootstrap site colors but don't adjust the layout.

3. d. So you don't have to make any big changes to your site theme.

4. False. You should always test any theme you install as extensively as you can.

5. b. It is easy and quick to do. Building a WordPress theme takes a long time and a lot of work even if you use Bootstrap to speed up the process.

6. True. You can find both online and offline Bootstrap editors.

7. d. Use a prebuilt theme or template. This is the fastest way to change the look of a Bootstrap website.

8. a. There is no difference between a plugin and an add-on. They are two ways of saying the same thing.

9. True. Thousands of people are using Bootstrap to build and maintain websites all around the world.

10. False. Bootstrap is a good start, but you can design nearly anything you can imagine using Bootstrap.

Exercises

1. Find a web editor that supports Bootstrap that you like to use. Test it using it on your website. What features do you like? What features do you dislike?

2. Pick out a Bootstrap theme or plugin that you find interesting, and see whether you can install it on your website.

3. Share your Bootstrap website on my Facebook page https://www.facebook.com/JenniferKyrnin. Feel free to ask any questions about Bootstrap or web design as well.

Index

U-V

W

X-Y-Z